The Representation of Science and Scientists on Postage Stamps

A science communication study

The Representation of Science and Scientists on Postage Stamps

A science communication study

Christopher B. Yardley, PhD

PRESS

Published by ANU Press
The Australian National University
Canberra ACT 0200, Australia
Email: anupress@anu.edu.au
This title is also available online at http://press.anu.edu.au

National Library of Australia Cataloguing-in-Publication entry

Creator: Yardley, Christopher B., author.

Title: The representation of science and scientists on postage stamps : a science communication study / Christopher B. Yardley.

ISBN: 9781925021783 (paperback) 9781925021806 (ebook)

Subjects: Science on postage stamps.
Scientists--On postage stamps
Postage stamps as propaganda
Postage stamps--Social aspects
Postage stamps--Political aspects
Postage stamp design--History
Visualization
Political geography

Dewey Number: 769.56

All rights reserved. No part of this publication may be reproduced, stored in a retrieval system or transmitted in any form or by any means, electronic, mechanical, photocopying or otherwise, without the prior permission of the publisher.

Cover design and layout by ANU Press

Cover image: The stamp on the cover was issued by Russia (USSR) in 1969 to celebrate the Centenary of Dmitri Mendeleev's Periodic Law. The message is conveyed through the portrait which dominates the image and reinforces the perception of the scientist as an older man, deep in thought at his desk as he ponders the periodic table. The context is well thought out, Mendeleev holds a pencil, and some of his workings are shown with chemical laboratory equipment in the immediate foreground. The level of context is high for a stamp issued in the 1960s.

This edition © 2015 ANU Press

Contents

Acknowledgments vii

1. Why Study Science on Stamps? 1
2. Developing a Taxonomic Structure 13
3. Classification and Analysis 39
4. Stamps as Communicators of Public Policy 77
5. On Being First 143
6. Scientists on Stamps 175
7. Two Time Capsules 225
8. Discussion 265
9. Conclusions 285

Bibliography 291

Acknowledgments

I am indebted to The Australian National University (ANU), particularly the Australian National Centre for the Public Awareness of Science (CPAS), for accepting me as a mature student after 45 years in business, selling computer systems. Upon the completion of my Master of Science Communication degree I was given the opportunity to undertake the research that resulted in this book. As a lifelong stamp collector I have built upon this knowledge to build and study the relevant stamp database.

I particularly acknowledge the support and encouragement of Professors Sue Stocklmayer AM (Director of CPAS), Chris Bryant AM, and John Rayner AM.

ANU Press has been most helpful in guiding me through the publishing experience and has also afforded me a subsidy to help with copyediting costs.

Throughout my working life I have enjoyed friendship and support within teams with an agreed purpose. ANU has been no different and I have enjoyed the experience.

My wife, Audrey, was not initially keen for me to make the commitment to complete a PhD. That said, she has provided a total support and we are looking forward to my coming 'gap-year' and a start on the list of jobs I have been putting off for more than four years.

Christopher B. Yardley
2015

1. Why Study Science on Stamps?

> Postage stamps are a very political, territorially grounded and yet overlooked part of visual culture (Raento and Brunn, 2005, p. 9).

For approaching two centuries, the images on postage stamps have been used to convey messages from the government of the day to the general public. Science has been used to enhance those messages for the past nine decades. In this book, I explore the ways in which science and scientists have been portrayed on stamps and look at the ideas and, in some cases, the propaganda that underpins them.

Communicating science through stamps has attracted few examples of scholarly analysis. Various people have looked at specific scientific disciplines, such as chemistry or medicine, on stamps, but a sweeping overview such as is given in this book has not been previously conducted. The idea that stamps are a scholarly topic is not, however, new. Child (2008), for example, argues that the fact that philately is generally regarded as a hobby does not preclude study. He maintains that most of science and the humanities started as hobbies; as culture changes, the value of intellectual pursuits also changes. He also points out that recent scholarly work dealing with popular culture has included postage stamps and even comic books, focusing upon their contributions to the history and politics of the age and of the area being considered. As early as 1969, Ekker wrote: "stamps, as government documents with important content, should be accepted by scholars as legitimate primary source materials for research purposes" (Ekker, 1969, p. 40).

The public's relationship with science is, increasingly, the subject of study within the discipline of science communication. Since the late 1980s, the focus has been on the public understanding of science (PUS). From the mid-1990s to the present, there has been an increasing critique of PUS and the development of the public awareness of science (PAS), and the public engagement with science and technology (PEST). The postal administrations of the world, in using science as a medium for communicating specialist knowledge and experience, through their messages act as mediators between ambition and reality and their published view at the time of issue which will change over time. These changes, which reflect the changes in perceptions about science communication, are also examined in the book.

The stamp as a communication device

Postage stamps are issued in huge numbers. In Australia today, for example, the print run for a commemorative issue stamp will be at least 200,000 copies for gummed stamps, and possibly as high as 2 million for the every-day postal delivery charge, especially when sold for business purposes in sheets and rolls. Numbers several times larger than the Australian print runs will be appropriate for countries such as Russia and the United States. The US Postal Service, for example, prints 50 million stamps as an initial print run for some commemoratives (Linn's Stamp News, 17 April 2012). Postage stamps are distributed to (and by) the general public across the country of issue and the whole world. The message contained within the stamp has a lifespan that lasts for as long as an example of the item is contained in a collection.

> Stamps make up part of what the Australian political scientist has termed "the public culture", namely a set of images and values which are propagated as the taken-for-granted picture of the world (Altman, 2010, p. 3).

Former United Nations Secretary General, Javier Perez de Cuellar, assessing the role that postage stamps play in achieving the United Nations' goals, stated: "Stamps are a form of communication and culture. They carry a message of their own and lead to world understanding" (Child, 2008, p. 12). The same cannot be said of most of the media, and postage stamps are undoubtedly part of the media. Just as newspapers and electronic media are analysed for the meaning and impact of their messages, so too stamps offer an opportunity for this kind of analysis. This book represents the first phase of analysing the representation of science on stamps: the number, nature and meaning of the messages.

Postage stamps provide a means of communication between the issuing authority and the members of the general public who buy and use them. During the passage of mail, several people may be exposed to the message contained in the narrative of the stamp, be it visual, textual, or a combination of both. For more than 170 years, the issuing authority has been an actual agency of government. Today, in the twenty-first century, the postal authority will more likely be a state-owned enterprise reporting to a minister of government, and is still seen as an official medium for the dissemination of messages from government. Since there is only a short window of exposure, the message must be carefully crafted. Thus, any science messages not only represent a government view, but will also inform about the way in which science itself is viewed at that time, in that place. Michael Zsolt of Australia Post envisages a three-second window of engagement to grab the attention of someone handling an envelope carrying an Australia Post stamp (Zsolt, 2012). Clearly, both the design and the science must instantly captivate if the stamp is to be noticed at all.

The scope of this book

I cannot remember when I have not been a stamp collector, or perhaps a better description might be a stamp accumulator. Although I have comprehensive accumulations of stamps from Australia, New Zealand Great Britain and Ireland, four of five countries in which I have lived and worked and which feature in this book, one stamp in particular prompted my quest to understand science on stamps. The stamp is shown as Figure 1.1.

Figure 1.1: China, 1980. *The Second National Conference of the Chinese Scientific and Technical Association.* **Gibbons catalogue # 2974.**
Source: Author's collection.

The stamp was issued by the People's Republic of China in 1980, (the date is shown at bottom right). But what are female forms in diaphanous gowns, surrounded by scientific images, doing on a postage stamp just four years after Mao Zedong's Cultural Revolution? It is questions such as this that intrigue and engage, and prompt further analysis.

The ten countries whose science stamps are considered in this book are Australia, New Zealand, Great Britain, France, Germany, Ireland, Poland, Russia, China and the United States of America. These countries are not a random choice. They have been chosen to reflect a selection of the influential and representative countries of the world, including those perceived to be of special interest due to their changing political situation and how this has been reflected on stamps. The political situation is discussed vis-à-vis stamp issues in the separate country reviews that are included within this book.

Although there has been some work done towards recognising the messages contained on postage stamps (see, for example, Altman, 1991; Scott, 1995), the study to date has been largely focused upon countries or regions, the development of icons, or as part of the history of specific scientific disciplines. Furukawa (1994) took an extensive look at medical history and has written a comprehensive review of the achievements of each doctor who has been celebrated with an appropriate image on a postage stamp. Weber's *Physics on Stamps* (1980) takes a slightly different approach, with chapters on topics such as 'Equations' and 'Systems of Measurement'. Wilson's *Stamping Through Mathematics* (2001) provides a short biography alongside approximately 400 stamps showing mathematics or a mathematician. Journal articles tend to follow this format, as do the deliberations of particular thematic collecting groups such as the Biophilately Group and the Mathematical Study Unit of the American Philatelic Society. Such deliberations are strictly focused. No one has yet looked at stamps in an effort to match up the science of the day, the choice of stamp image and the messages contained within the stamp.

De Young observes that "increasing realisation of the mutual interdependence of the scientific community and national development is reflected in the fact that postage stamps directly utilising themes based on science in their illustrations are exceedingly rare before 1950, but appear with increasing frequency thereafter. Science is too important to national progress for it to be ignored in government policies" (De Young, 1986, p. 1).

Both science and scientists provide the subject matter for many of the world's postage stamps, numbering as many as 30,000 over the past 80 years. Kevane has written: "Quantitative measurement of the imagery on postage stamps has several virtues. Such quantification permits a subtle, complex and continuous measurement of regime strategy, since typically numerous stamps are issued in every year" (Kevane, 2006, p. 4).

Quantitative measurement is one aspect of analysis. Qualitative analysis, as advocated by Scott (1995), in particular the analysis of semiotic considerations, also offers insights into the science depicted on stamps. Through the use of signs, humans represent ideas, ideals, objects and philosophy. In my analysis, semiotics offers a means of studying the qualitative role of signs in human culture and social interaction as represented by the portrayal of science on stamps. There has never been a study of *how* science on stamps conveys the image of science, and whether it mirrors contemporary thinking about science and how it should be communicated.

This book sets out to answer the challenge of the authors cited above who have encouraged closer study of the meanings inherent on postage stamps.

1. Why Study Science on Stamps?

That challenge will use the images of science and scientists from the range of countries, including those that have been through different political structures in the 170 years of the life of the postage stamp.

A major question which I set out to answer is: what does the representation of science and scientists on postage stamps convey about the political and cultural necessities of a country at the time of issue?

For some countries, the importance of historical claims has led to the issue of stamps about 'being first'—the importance attached to the date of a scientific discovery and the world's acknowledgement of the achievement (see Chapter Five)—or the issue of stamps that reflect particularly important discoveries on their own soil. For these countries, the history of science is a matter of national pride.

China has, perhaps, set the precedent for a set of stamps with a historical bias that invites comparison with modern scientific instruments that would be generally be familiar to the public. The Chinese 1953 set, presented in Figure 1.2, shows (from left to right): a compass from the 3rd Century BC; a seismoscope, to record earthquakes, from the Han Dynasty (132 CE); a drum cart, to measure distance, from the Chin Dynasty (300 CE); and an armillary sphere, modeling objects in the sky, from the Ming Dynasty (1437 CE).

Figure 1.2: China, 1953. *Major inventions by Ancient Chinese scientists.* Gibbons catalogue # 1601–1604.
Source: Author's collection.

China has acknowledged the scientists of Ancient China with a number of sets. One such set, released two years after the major inventions issue, is shown in Figure 1.3.

Figure 1.3: China, 1955. *Scientists of Ancient China*. Gibbons catalogue # 1660–1663.

The scientists are (from left to right): astronomer Cheng-Heng (78–139 CE); mathematician Tsu Chung-chi (429–500 CE); astronomer Chang-Sui (683–727 CE); and pharmacologist Li Shih-chen (1518–1593 CE). Source: Author's collection.

In contrast, Ireland does not issue very many stamps and is, in fact, the smallest contributor to this study, but I do appreciate their issue of 1981, titled *Irish science and technology* (Figure 1.4). Four stamps were issued that shout to the world: "Please note that these Irishmen have contributed to world science and technology, although their achievements may have been used and acclaimed by other countries." The stamps are tied together by a common design. The message is one of nation-building and civic education, and is political. The scientists' inventions float above their portraits rather like the thought-bubbles of comics.

Figure 1.4: Ireland, 1981. *Irish science and technology.* Hibernian catalogue # C30–307.

12p: Robert Boyle (1627–1691) was an Irish-born chemist, physicist, inventor, and early gentleman scientist, noted for his work in physics and chemistry. Great Britain has also claimed Boyle as a local scientist, albeit one from a colony.

15p: Harry Ferguson (1884–1960) patented the automatic control system, which is now employed by almost all tractor manufacturers worldwide, in 1925.

16p: Charles A Parsons (1854-1931), invented the modern manifestation of the steam turbine, a mechanical device that extracts thermal energy from pressurised steam, and converts it into rotary motion, in 1884.

25p: John Philip Holland (1840–1914) was the engineer who developed the first submarine to be formally commissioned by the US Navy. He has also been celebrated on the stamps of Great Britain.

Source: Author's collection.

Australia has been quite late to join the party, publishing the *Australian Achievements in Technology* set of stamps in 1987 (Figure 1.5). The title of the issue is included on each stamp, as is a short name describing the icon shown. The value of the stamp is shown as an LED number, emphasising the technological aspect of the message.

Figure 1.5: Australia, 1987. *Australian achievements in technology.* **Renniks catalogue # 1024–1027.**
Source: Author's collection.

The names of products and manufacturers featured are (from left to right): Bionic ear (University of Melbourne, Otolaryngology Department); Microchips (Austek Microsystems Pty. Limited, Adelaide); Robotics (Machine Dynamis Pty. Limited, Melbourne); and Ceramics (the image showing a nuclear magnetic resonance spinner). This is the only stamp issue I know of where the name of the manufacturer is publicly known and has been published in the local stamp catalogue. It takes the Australian invention to the world. It is an advertisement on a postage stamp, emphasising the fact that Australia is innovative and 'open for business'. It openly raises awareness of science and technology.

Chapters Five and Six examine not only claims of being first, but also the creation of scientific heroes through their images. An important question I shall attempt to address is whether the representation of science and the implied message on the stamp has reflected changes in the public comprehension of science over the past 90 years. As science and technology change the way the general public lives and works, it can be anticipated that the messages might reflect changes in science and technology, the way these are presented, and what they might mean to the public. The debates surrounding science communication have developed through the phases of the public understanding of science to embrace awareness and engagement of the public with science.

A second aspect of this analysis therefore asks two questions. Firstly, is the public awareness and perception of science reflected on postage stamps? And secondly, have stamps been issued that contribute to the public awareness of science? Is there, for example, any major difference in the communication of science between stamps depicting the life and achievements of, for example, Darwin and Einstein.

This book explores how and why science and scientists are chosen to represent what a particular country wishes to publicise as its aims, ideals and ambitions both internally, within the country, and to the outside world. The book contributes to current theories on science communication, and will also inform historians. As part of the analysis of science communication, I include a taxonomy for evaluating postage stamp images that may be extended to other visual media such as advertisements. The taxonomy is not restricted to an evaluation of the representation of science contained in a stamp message, as it can also be applied to other aspects, such as historical or cultural concepts. In looking at each image, I have also asked the question: is this a mirror (of reality), or is it a lens? This question permeates every aspect of the analysis.

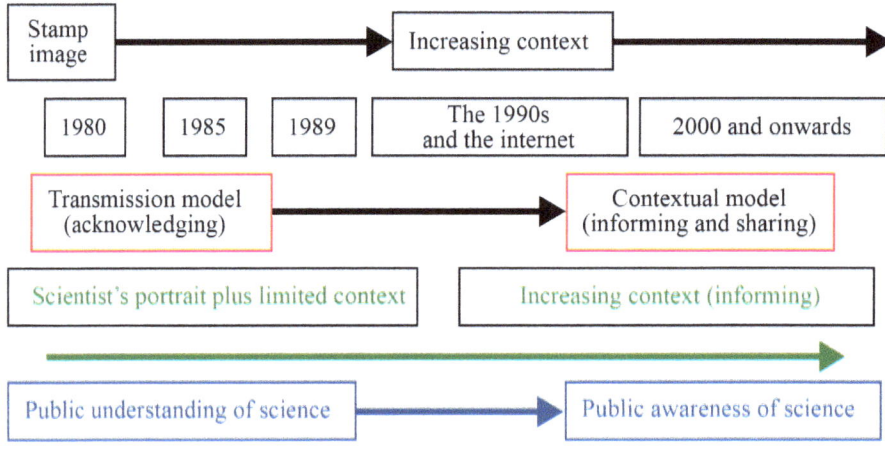

Figure 1.6: The model developed in this book.
Source: Author's research.

The stamps studied

What is a postage stamp?

From its introduction in Great Britain in 1840, the postage stamp's role has been as a fiscal device. It pre-pays a service to be performed by the dispatch of an article through the world's postal services. The postage stamp is not intended to be in the province of the privileged few. It is printed for everyday use and its messages are conveyed to the world, without prejudice, through a mediator who

1. Why Study Science on Stamps?

knows the audience and tailors the communication to meet defined objectives. The function and physical attributes of the stamp are little changed with time, except, perhaps, for the recent nicety of self-adhesive stamps rather than gummed version, which necessarily eliminates the possibility of a DNA identity clue, beloved of television and the detective novel.

Quite simply, in physical terms, the ubiquitous postage stamp is a small piece of gummed paper that is purchased from a post office and attached to an item of mail as evidence of pre-payment of the fee for delivery to an address also displayed on the item. Typically, stamps are printed on special paper, with a national designation and denomination (price) on the face, and an adhesive on the reverse side.

Figure 1.7: The main components of a stamp.
Source: Author's research.

The stamp shown in Figure 1.7 is from a 1995 set of four Australian stamps celebrating medical science. Two scientists are portrayed: Dame Jean Macnamara (1899–1968), best known for her contributions to children's health and welfare, and Sir Frank Macfarlane Burnet (1899–1985), an Australian immunologist and 1960 Nobel laureate. The image shows a pair of hands preparing an injection against a background suggesting medical research. It is a good example of a mid-1990s image showing the scientist and the context through which they are recognised.

The unused postage stamp has an inherent monetary value that can, within reason, be redeemed through use at any time. What makes the postage stamp unique is that the (philatelic) collector market has given even the used item a value, especially if the cancellation on the stamp, illustrating that the service has been provided, gives the stamp a unique history by stating, in ideal circumstances, the date and place of service fulfillment.

Postage stamps may be printed in millions of copies and a certain number will be put into collections in their unused or used condition. Once in a collection, the stamp has an almost perpetual life and a continuing value. One only has to look at postage stamp auction lists, published regularly, to see that a collection maintains a personality and a life of its own.

The issuing authority prints new stamp designs on a regular basis, almost monthly, to pre-pay its most commonly used services, with the unused stamp retaining its service-value forever. Great Britain, for example, issues stamps that are designated 'NVI 1st' (No Value Indicated, first-class) which pre-pay for a first-class service delivery whenever presented, regardless of the cost of the service. The United States has recently followed that lead and is issuing stamps known as 'forever stamps'.

Stamps are issued as either definitive or commemorative. Definitive stamps are issued over a range of prices, are generally designed in a constrained, conservative style and will be available from the post office for a number of years. In Australia, definitive stamps are available in price ranges from five cents to ten dollars in order to enable post office clerks to provide a bespoke pre-payment postage of almost any value. The themes of the message on definitive stamp are country specific. Great Britain, for example, has used the same design on its definitive stamps, the ageless image of the monarch, Queen Elizabeth II, for more than 40 years. Australia tends to present illustrations of native flora and fauna on its definitive stamps. Commemorative stamps, on the other hand, are issued with a limited number of values, those for the most requested services, and follow a specific theme. About half of all commemorative stamps record a particular anniversary or event.

Figure 1.8: Examples of typical stamp formats/images described in the text.
From left to right: The Great Britain standard definitive format, an image unchanged since 1967, with the denomination '1st' (first-class internal mail service); A typical Australian fauna definitive stamp; Examples of typical Australian anniversary-driven commemorative stamps. These last two stamps are from a set of four, commemorating *Inland Explorers*.
Source: Author's collection.

The relevance of the postage stamp today

A major question that is often raised nowadays is that of the viability of postage stamps when postal authorities are facing a constant decrease in the volume of mail. Big business is now reluctant to use postal services for such previously printed documents as bank statements, annual reports and formal announcements. Such changing practices (often disguised as green initiatives), it is argued, mean less use of the mail system. Indeed, even a cursory look at the mail in any post box will show that fewer envelopes carry a printed postage stamp. Many businesses have, for years, used an in-house automatic franking and accounting mechanism, which obtains a discounted postage rate. The business will also have added a message of its own (this practice subtly emphasising the idea of the stamp as a viable carrier of messages). Plain paper receipts are now often used for parcels. So will stamps die? The consensus is that they will not. The stamp dealership community is declaring an increased interest in the classic stamps of the nineteenth century and in examples of material that has genuinely passed through the postal system and been used for the purpose it was intended. The UK *Telegraph* (7 December 2009), citing several sources, reports that stamp "dealers are reporting a sharp increase in sales" (Liew, 2009). The main argument by many dealers is that too many stamps are being issued, with the inherent dual danger of the general public and also stamp collectors losing interest and, in the longer term, stamps losing their appeal as investments. Jennings (2012) wrote in the *Gibbons Stamp Monthly* of February 2010 that major stamp dealerships in the UK have cancelled their long standing orders for new issues from Royal Mail because of declining sales as their customers become disenchanted with so many new issues. Paterson (2009), however, summarised the situation as he saw it:

> Will postal services survive? In my opinion survival is not only likely but virtually assured; even though the conveyance of documents, small purchases, gifts and written communications, cards and so forth may become highly unprofitable in the future. It is difficult to imagine modern civilization stripped of its ability to make these sorts of transfers between individuals (Paterson, 2009, pp. 5–6).

Australia Post has reported that mail and courier services are being expanded to cope with the increase in parcel deliveries as a result of internet sales (Australia Post, 2011, p. 3).

In addition to the continuing use of casual post office purchase of stamps there is a concerted effort by the issuing authorities "to protect and grow the philatelic market":

Stamps are profitable – the stamps not used for mail are collected.

There are millions of collectors worldwide (~30M according to the Universal Postal Union).

Stamps are a tangible representation of the Royal Mail brand.

> Stamps have the ability to touch a range of markets and generate positive PR.
>
> (Parker, 2004).

More stamps are being issued in the twenty-first century than ever before, with a greater range of subjects, including science and science-related themes.

I hope this book will also be of interest and value in a local context across disciplinary boundaries for the issuing authorities and my fellow philatelists, widening the current understanding of classification by themes and enabling further study of the scientific aspects of stamps in countries other than those in this study.

2. Developing a Taxonomic Structure

> However, even when the stamp represents an aspect of a country, such as a monument or site, it equally represents the country itself, for the image reproduced on the stamp is accompanied by the signs which establish that nation's identity (Scott, 1995, p. 8).

Bernard Smith, doyen of Australian art historians, is quoted as having said "an image is information" and that in assessing an image it should be "tested for validity and intention" (Palmer, 2012). The messages being studied in this book are those contained within postage stamps representing science, which I define later in this chapter.

As issued by the postal authority, stamps fall into two main categories. The first is the definitive issue, general purpose stamps which are on sale, usually for a period of a few years, over the post office counter. These are sold to pre-pay a postal service over a wide range of prices to enable the user to buy a product to meet his or her specific need. The second class of stamp, the commemorative stamp, generally has a specific themed image and will have a limited shelf-life in the post office. Commemorative stamps, also known as special stamps, are issued as a single stamp celebrating a theme or significant achievement, or are combined in a set of stamps at the prevailing prices for letter post to popular designated destinations.

The themes that are addressed on commemorative stamps can be very general, or quite specific, if, for example, medical scientists are the chosen subject. Commemorative stamps are issued to communicate a specific message for a limited time. This study is concerned with understanding scientific content messages, how and why themes are chosen, and the timing of the issue. The vehicle conveying the message might be a definitive or a commemorative stamp.

The over 200 postal authorities of the world have, to date, printed some 600,000 different stamps. Perhaps as many as 60,000 (10%) of these will have a representation of a scientist or science as a part of the image. The first portrayals of science on stamps, according to my below definition of science, did not occur until the 1920s and 1930s, after which time issuing authorities have celebrated scientific and technical achievements to promote government policy within their own country boundaries and, in turn, to the world.

Essential in determining the database for my study, I have used the listings of all legitimately printed postage stamps recognised by four reputable, international stamp dealers who provide whole-world stamp listings in their catalogues. Additionally, most countries are serviced by three or four local stamp dealers who list that country's postage stamps as their sales catalogue. Two or more catalogues have been used for each country considered to ensure completeness and verification of data. Catalogues have been used as data sources and to provide a reference to specific stamps discussed within the text.

Some countries are accused of printing stamps solely for the collector and their issues may not even be sold in the post offices of that country. From 2002, the Universal Postal Union (UPU) has published online the images and catalogue numbers of the stamps of the participating countries that are registered with them. These stamps must meet the legitimacy criteria of having been readily available in the country of issue and sold at face-value at the time of issue for inclusion on the website. My study is limited to science on stamps of countries that have fulfilled these criteria. However, I also look beyond these strictly legitimate boundaries to include the stamp issues of four of the Antarctic Territories, stamps that Altman (1991) suggests are only issued to justify the controlling country's possession and occupation of disputed land.

By and large, commercial catalogues number the stamps issued by a country's postal authority from the first issue, in date order, and increment the subsequent issue chronologically, as does the UPU World Numbering System (WNS). This simplifies the count of the number of stamps produced by a country and specifies the time-frame. In some cases here, when identifying a particular stamp, the UPU WNS is used if the stamp issue is too recent to have been catalogued in a vendor's hard-copy publication.

It should be a simple matter, therefore, to look at the most recent stamp catalogue for a country to determine how many stamps that country has issued. It is noted here that the number of stamps being issued annually has been higher in recent years than the average number of issues over time. No distinction is made here as to how a stamp has been sold, because in most instances this cannot be determined by examination. The postal authority might have sold the stamp as a single from a sheet, within a miniature sheet or sold via a booklet over the post office counter or through a vending machine. Sweden, for example has sold the majority of its stamp issues in booklet or coil via vending machines since the 1920s. ('Nimrod', 2001).

Defining science on stamps

I could, initially, find no convenient definition of science so I have developed my own. I have included most of the disciplines that are included in a university science and applied science curriculum. Biology, botany and zoology are included, but stamps that depict flora and fauna and dinosaurs are excluded, even if they are described with their scientific names. I adjudge these subjects as travel and tourism advertisements for the country and collectors of these themes, rather than to send a scientific message. However, if a stamp or a set of stamps is dedicated to the celebration of a named scientist with flora and fauna shown to add a contextual description of his or her achievements, they are judged to be scientific. This is further discussed in Chapter Three. Archaeology and architecture are not included; geology, computing technology and astronomy are. Engineering is included following a similar argument to that for flora and fauna: if engineer designer is the main focus of the issue, it will be included. This latter category has been important for the Russian series of airplane designers and other technologies that have been featured, over the years, with images showing the development of their designs.

Part-way through my study, the Australian Government Department of Innovation, Industry, Science and Research issued its 'Inspiring Australia' initiative. In this strategy it defined science or the sciences to be:

- the natural and physical sciences such as biology, physics, chemistry and geology
- the applied sciences, such as engineering, medicine and technology
- newly emerging and interdisciplinary fields, such as environmental science, nanotechnology and phenomics
- mathematics, a field of study in its own right, as well as an essential tool of the sciences
- the social sciences and humanities, critical to the interface between science and society. (Department of Industry, 2010, p. ix).

This definition is consistent with the approach I had been taking. It is relevant to see the social sciences and humanities included, but in this study they are only included if they constitute part of a directly scientific message that is being told through the postage stamp requiring a science understanding: "For example, science-related areas include health promotion, science teaching, nursing, agriculture, science and environmental policy development, etc." (Department of Industry, 2010, p. ix).

I contend that postage stamps are a part of the media, or behave as though they are. In many ways they regularly publish stories that will be of general interest to the public. The messages on the stamps are someone's idea of a story worth telling at the time. One element of my study has been to determine the extent to which government is sponsoring the message to be told. Some attempts have been made to define science from a media perspective and I have considered several precedents. Lemhkuhl *et al*. (2011) suggest that classifying a media subject as science might be based on the following categorisation: "Science is the real subject matter; scientific insights are used to explain everyday phenomena or social problems, or offer orientation in a complex situation in which science appears as a service provider; or, scientific method is the focus that problematises scientific findings or the pursuit of scientific findings." (Based upon Lehmkuhl *et al*, 2011, p. 8)

These classifications are very much science-centric dependent upon a type of science. The general public, however, does not consist of a specialised science audience. I believe this categorisation is, therefore, not applicable because the messages to be sent will be almost stand-alone and not precisely structured. It also has to be remembered that the message might be contained within a single stamp, or as a story told within a set of stamps or a series over time, and who but the stamp collector is likely to see the set in its entirety?

A few previous attempts have been made to classify the images used to represent science on postage stamps. In his paper "Postage Stamps and the Popular Iconography of Science", Gregg De Young described his classification as: (1) images of specific scientists; (2) images of scientific workers in general; (3) images of scientific research institutions; (4) images of scientific equipment; (5) images of natural phenomena; and (6) other miscellaneous symbols, such as scientific formulae (De Young, 1986, p. 3). De Young describes what he means by each category and offers examples of each. He studies the use of scientific equipment as images to represent science to the general public, concluding that these devices ignore the intellectual or the theoretical activity that is the cornerstone of science.

De Young also argues that representing science by showing the tools of science ignores the diligence, scholarship and knowledge of the process that constitutes science. But he concludes it is official government status that gives "these illustrations an authority in shaping the popular view of scientific activity that far exceeds the limits of their popular size" (1986, p. 13). This limitation is overcome in this study by confirming and making note of the postal authority's description and reason for the issue as described in the postal authority's publications and the relevant stamp catalogues.

Another limitation in De Young's classification is that he has made no attempt to determine the timing of the stamp issue. This is relevant as very many issues, particularly later issues, are directly associated with an event or the anniversary of an event, the anniversary of the scientist or of their achievement for example. For my study it is important to recognise science that has been institutionalised and presented by politicians for general consumption and political ends, such as the climate change science.

The representation of science and scientists

The science stamps examined fell into the classifications of the dictionary definitions of representation shown in Figure 2.1.

> representation
>
> noun
>
> 1a. An image, likeness or reproduction of a thing; spec. a reproduction in some material or tangible form, as a drawing or painting,
>
> 1b. The action or fact of exhibiting or producing in some visible image or form,
>
> 1c. The fact of expressing or denoting by means of a figure or symbol; symbolic action,
>
> 2a. The action of presenting a fact etc. before another or others; an account, *esp*, one intended to convey a particular view and to influence opinion or action,
>
> 2b. A formal and serious statement of facts, reasons or arguments made with the aim of influencing action, conduct, etc,; a remonstrance, a protest, an expostulation,
>
> 5a. The action of presenting to the mind or imagination; an image or idea thus presented.

Figure 2.1: A dictionary definition of the word "representation".
Source: Shorter Oxford Dictionary, 2010, p. 2553.

Countries chosen for study

Ten countries have been chosen for study. These countries represent different political ideologies and include changes of national ambition over time. The counties range from the largest in terms of population, (China, Russia, USA), to smaller countries (Ireland and New Zealand). Some of the countries are long established, (Great Britain and France), while others have been through political turmoil, (China, Germany and Poland), since the introduction of postage stamps. Russian stamps, numerically, constitute some 40% of the science stamps identified. During the 1960s and 1970s, Russia actually issued one in five of all stamps issued world-wide (Mackay, 1976). The ten countries included in my study are:

Australasia

1. Australia: A stable democracy. Federated into a Commonwealth by public vote from six existing (British) colonies on 1 January, 1901. It was 1913 before the first federation stamp was issued. Explorer Captain James Cook had been featured in the world's first commemorative set of stamps by the colony of New South Wales to celebrate the centenary of the First Fleet in 1898 and this stamp is included in my study. This study includes the postage stamp issues of Australian Dependencies: Norfolk Island, The Australian Antarctic Territory, Christmas Island, and the Cocos (Keeling) Islands where relevant.

2. New Zealand: Australia's nearest neighbour and another stable democracy. The first New Zealand stamp was issued in 1855. My study also makes mention of the stamp issues of the Ross Dependency as the Antarctic location of New Zealand research interests where relevant.

Europe

3. Great Britain: A stable democracy. Stamps issued from 6 May 1840 to the present day. The stamps of the British Antarctic Territory will also be examined.

4. France:

   a) Second Republic, 1848–1852.
   b) Second Empire, 1853–1940.
   c) French State during 1940–1944.
   d) Provisional Government, 1944–1946.
   e) Fourth Republic, 1946–1958.
   f) Fifth Republic, 1958–present.

The stamps of the French Antarctic Territory will also be examined.

5. Germany:

   a) Empire, 1871–1918.
   b) The Weimar Republic, 1918–1933.
   c) Third Reich, 1933–1945.

   Allied Occupation

   a) Allied Military Post, 1945–1946.
   b) American, British and Soviet Russian Zones, 1946–1948.
   c) British and American Zones, 1948–1949.

   d) French Zone, 1946–1946.
   e) Russian Zone, 1945–1949.

   German Federal Republic

a) West Germany, 1949–1990.
b) West Berlin, 1948–1990.
c) Germany reunified, 1990–present.

German Democratic Republic

a) East Germany, 1949–1990.

6. Ireland: The stamps of Great Britain were used until 1922.

   a) Irish Free State, 1922–1937, Great Britain stamps with overprint and state issues.
   b) Republic of Ireland, Ēire, 1937–present.

7. Poland:

   a) Russian Province, 1832–1918.
   b) Republic, 1918–1939.
   c) German Occupation, 1939–1945; Polish exiled government in London 1941–1944.
   d) Republic, 1944–1948, a Soviet-model communist state.
   e) Social Democracy of the Republic of Poland, 1990–present.

8. Russia:

   a) Empire, 1853–1917.
   b) Russian Soviet Federative Socialist Republic, 1917–1922.
   c) Union of Soviet Socialist Republics, 1922–1991.
   d) The Russian Federation, 1991–present.

Asia

9. China:

   a) Monarchy, 1850–1949.
   b) The People's Republic of China, 1949–present.

The Americas

10. The United States of America. A democracy, its stamps issued from 1850 to the present day.

Analysis

The preliminary analysis examines every issue of a country and makes the subjective evaluation: Is this science? Does it celebrate the achievement of a specific scientist? Does it celebrate a scientific breakthrough? Is the issue representing a scientific or institutional message. Is the image reflecting science as a public service?

Sourcing stamp images

It is important that I am able to look at the stamps as they have been supplied to the general public, who are the potential recipients of the messages being sent on postage stamps. Starting out on this project, I had the advantage of being a long-term stamp collector and owning a specialist collection of three of the countries I am examining: Australia, New Zealand and Great Britain. I do have some stamps from the other seven countries, but my collection is far from comprehensive. I needed to acquire approximately 80% of the stamps that I have classified as containing a science message after a detailed scrutiny of stamp catalogues.

Figure 2.2 shows approximately one-eighth of a page of a typical Stanley Gibbons catalogue. At random, I have chosen to show a set of Russian stamps from 1963. It is a set that is of interest in the case study looking at the stamps of Antarctica, and the continuing scientific and research interest shown in the region.

You will notice that only one stamp from the set is illustrated, and that the three other stamps in this set are described as being of type 1041. I note here that the second image, type 1042, is not science, and therefore not of interest in my study. The other three stamps from the *Arctic and Antarctic research* set are described textually. When compiling my list of science stamps I am required to have and scan all four stamps in the set that comprise the total message for this issue.

2. Developing a Taxonomic Structure

1041 Antarctic Map and Supply Ship *Ob*
1042 Letters and Transport

(Des Ye. Aniskin. Litho)

1963 (16 Sept). Arctic and Antarctic Research. T **1041** and similar horiz designs inscr "1963". Multicoloured. P 12½×12.

2894	3k. Type **1041**	50	30
2895	4k. Convoy of snow tractors and map	60	30
2896	6k. Globe and aircraft at polar base	95	45
2897	12k. *Sovetskaya Ukraina* (whale factory ship), whale catcher and whale	1·90	85
2894/2897	Set of 4	3·50	1·70

(Des Yu. Ryakhovsky. Photo)

1963 (20 Sept). International Correspondence Week. P 11½.

2898	**1042**	4k. bluish violet, orange and black	40	15

Figure 2.2: Russia, 1963. Gibbons catalogue reference for the set of four stamps. *Arctic and Antarctic research*.

Source: Stanley Gibbons Catalogue, Part 10, Russia, p. 109.

Having compiled the list of the stamps to be included in my study, the next step was to acquire copies of the actual stamps. For each country I have a list of three or four dealers who might source the stamps. The function and practice of the stamp dealer is to be able to source against a collector's wants-list. Dealers typically purchase substantial collections and make their money breaking down the collection into parts to meet the requirements of purchasing customers. I have been fortunate to find dealers prepared to break down collections to be able to provide the science stamps I have requested. The two final columns from the Russia catalogue of Figure 2.2 are the Stanley Gibbons price for mint stamps (second column from the right) or used stamps (the right hand column) in British currency. Most dealers provide stamps against the Gibbons catalogue reference number but at a discount to the Gibbons price. I have managed to procure copies of almost all, (approximately 98%), of the stamps I needed to scan and include in my stamp image database. Those I have not been able to source are those that were either not, in the past, of interest to a general collector, or have become too expensive over time.

The stamps I believed I needed were not too difficult to procure and I used one main dealer per country. The only stamps with which I had difficulty were 1960s Australia, early Chinese issues and, surprisingly, the early Russian Federation stamps of 1992–1993. Dealers I approached for the latter stated that they had decided against supplying the new Russia stamps as they had not been able to dispose of the large numbers of stamps from the USSR period of Soviet Russia.

At the end of my project I shall have the opportunity to approach the same dealers to see if they want to buy back the science stamps I have acquired. It has to be said that none of the dealers to whom I suggested a loan, lease, or rental of the stamps involved were interested.

Construction of the taxonomy

A number of attempts have been made to classify science on some postage stamps. Jones (2001) makes observations about the images used to show particular sciences, and Webber (1980), Furukawa (1994) and Wilson (2001) have concentrated upon specific sciences. Kevane defines his categorisation based upon the kinds of stamps that are issued:

> Decisions have to be made when coding the imagery on stamps, even before arriving at a smallish number of categories to be used for grouping the wide variety of images and messages. These decisions arise from four sources of complexity in the kinds of stamps produced by countries: (1) some stamps are intended for the collectors market; (2) some stamps have different physical properties from the ordinary perforated, gummed stamps, properties that make them more or less suitable for use on letters; (3) the quantities produced and usage of stamps vary with the images on the stamps, in perhaps predictable ways; (4) and many stamps are issued in series rather than as stand-alone images. These four characteristics of stamps — collectability, properties, quantities, and series — have to be addressed in coding (Kevane, 2006, p. 5).

The Kevane coding regime has not been followed, because it will not help to answer the questions that are the focus of this study. For this study, a new scheme has been devised, as discussed below. Kevane also makes mention of the property of a stamp that might not make it suitable for use on letters. Very occasionally a post office has a pseudo-stamp issue that is not really suitable for everyday use. Such a device is not included within my study.

The taxonomy developed for my study overcomes the deficiencies noted above concerning previous definitions of science on stamps. For this project,

2. Developing a Taxonomic Structure

a subjective visual evaluation is made at the time of analysis. The conceptual representation of science and scientists on postage stamps is defined in the listing below.

The method involves a simple evaluation and count. The five primary categories are:

1. Is the stamp a single issue, with the one stamp relaying the message?
2. Is the stamp part of a set? Has a set been issued to tell a more complete story? If only one stamp in a set includes a representation of a science or a scientist, it is counted and noted as coming from a set.
3. Is the issue printed in a single colour or full colour?
4. Does the stamp name or recognise a particular celebrant as the main image of the design, the signifier and therefore the signified? As a symbolic process, what does the participant mean or represent? (Kress and Van Leeuwen, 2006).
5. Is the subject image non-personalised, showing the science, a scientific device, or the description of science in the service of the public to tell its message?

A schematic of the taxonomy used is shown as Figure 2.3 and is expanded with examples in Figures 2.4 and 2.5.

Another category that was considered is that of production of the actual image. Prevailing technology meant that the earliest stamps were printed in a single colour on white paper, although in times of shortage of white paper, papers of different colours were used as for expediency (and are a specialist delight for the avid collector). The first stamps printed with two or more colours were issued in 1887. Today's stamp designer occasionally uses a single colour as a device to represent a historical connotation to the image even when the complete palette of colours is available through the photogravure process. The early stamps were recess-printed from etched intaglio plates in much the same way as bank notes were produced. The initial prints were single colour, and it was only after WWII that two-colour intaglio became the norm. Lithography was used sparingly and eventually photogravure became the standard method. How the stamp is printed does not reflect the message except in rare cases that are best considered in isolation.

One example of where production is important, however, and which is a part of the story being told, is the Great Britain 2001 set of six stamps to celebrate Nobel Prizes and laureates. The stamp representing chemistry, for example, is a heat sensitive image of the carbon 60 molecule, celebrating the work of Sir Harold Kroto (born 1939), who was the Nobel Prize winner in 1996. Three stamps from this set are discussed in my study and are shown as Figure 3.11.

The taxonomic classification

My rationale for the taxonomy developed for the study has been explained. A schematic to illustrate the taxonomy is shown in Figure 2.3 and examples of the image classification are shown in Figures 2.4 and 2.5.

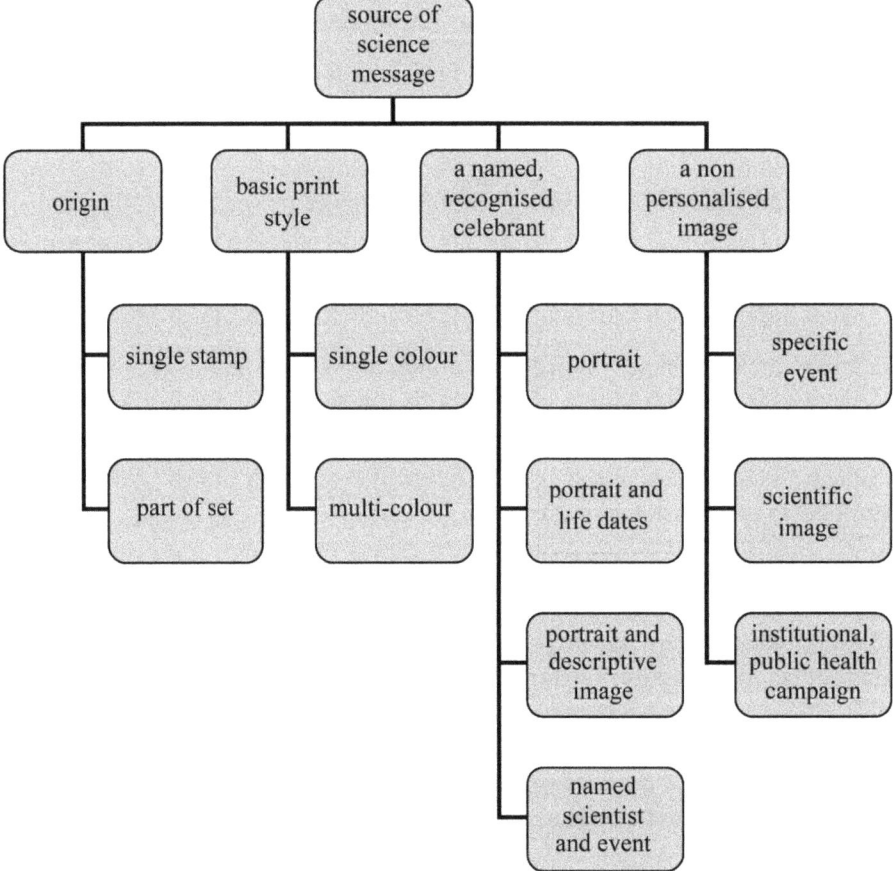

Figure 2.3: A basic taxonomy of the stamps studied, derived from Kress and Van Leeuwen (2006).

Source: Author's research.

In Figure 2.4, I show examples of the four classifications of the stamp images used in my study that have a named recognised celebrant. A narrative description of each classification follows:

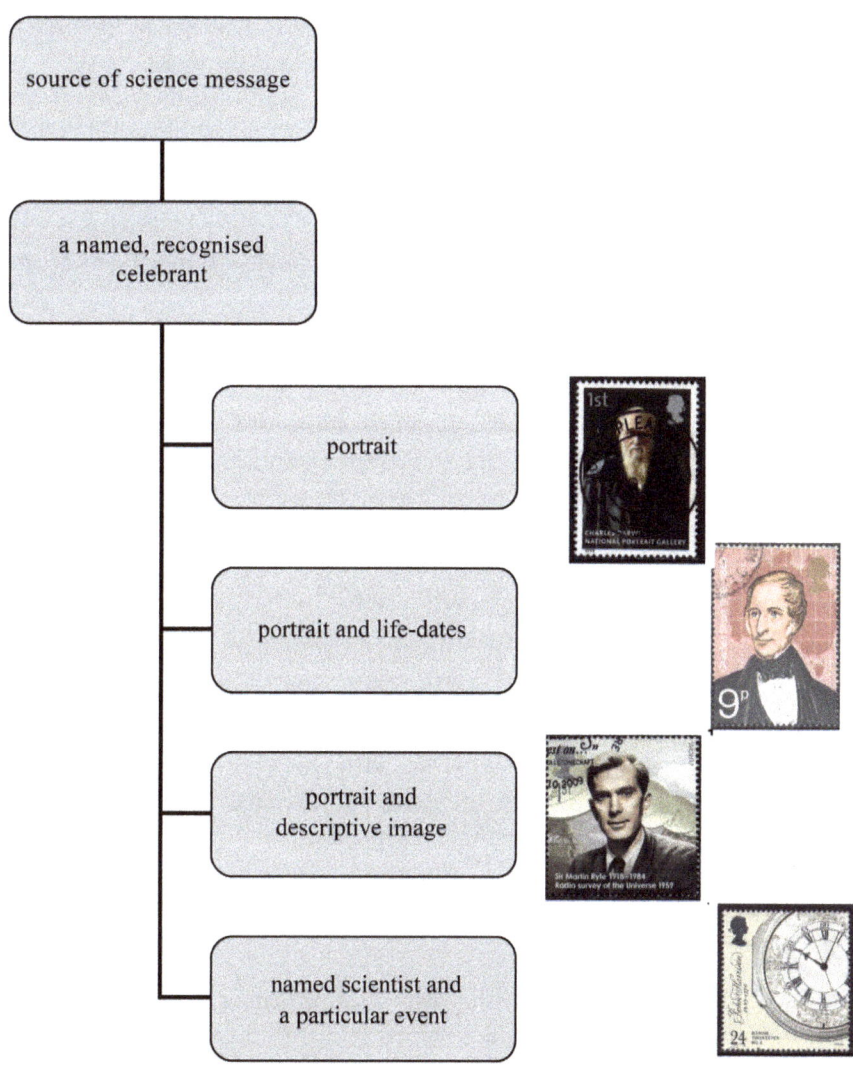

Figure 2.4: Examples of four taxonomy classifications of stamps that celebrate an identified scientist.

Source: Author's research.

Portrait: The portrait here is of Charles Darwin, from a 2006 issue commemorating the 150th anniversary of the National Portrait Gallery. He is the only scientist from a set of ten portraits of various well-known people painted by different artists. The portrait is the icon. The artist is John Collier (1850–1934). Seven men are depicted (through six paintings and one sculptured bust) and three women.

All stamps are of the same (first class) value for mail service within the UK, and all carry the monarch's profile in the top right hand corner to identity the country of origin. All images identify the celebrant and the artist.

Portrait with life-dates: Here I show a stamp that includes a descriptive image in addition to the portrait name and life-dates. Charles Sturt (1795–1869) was an explorer of Australia. The background map of Australia provides the context for inclusion in the set of five portraits entitled *British explorers*. Issued in 1973, the Sturt stamp is the highest value in the set and could be expected to be used on mail to Australia.

Portrait and descriptive image: From a set of ten *Eminent Britons* published in 2009, astronomer Sir Martin Ryle is shown against a background of communication equipment. His achievement is described as the 'Radio survey of the Universe, 1959'. Eight male and two female scientists are celebrated in the set. The set is tied together with a muted background, a profile of the monarch in the top left corner, and a further design consistency naming the person and their achievement in two lines at the bottom of the stamp. All ten stamps are of the same value, (first class post, internal delivery in the UK) so the expectation is that these stamps will be used within the UK, except for overseas collectors.

A named scientist and a particular event: The event is the 300th birth anniversary of John Harrison (1693–1776), the inventor of the maritime chronometer. The image is the decorated enamel dial of his H4 clock, not a portrait of the celebrant. The text includes Harrison's name, life dates and the fact that the clock-face is of his marine timekeeper number four. No portrait is shown but the scientist John Harrison is named through his signature, which is shown. The example reproduced is the lowest value in the set of four, for second class mail delivery in the UK.

In Figure 2.5, I show examples of the three classifications of the stamp images used in my study that do not have a named recognised celebrant. They show a science in abstract. A narrative description of each classification follows. Semiotically, the three stamps are similar to the first classification in that the monarch's profile identifies the issuing authority as Great Britain and shows the pre-payment service price.

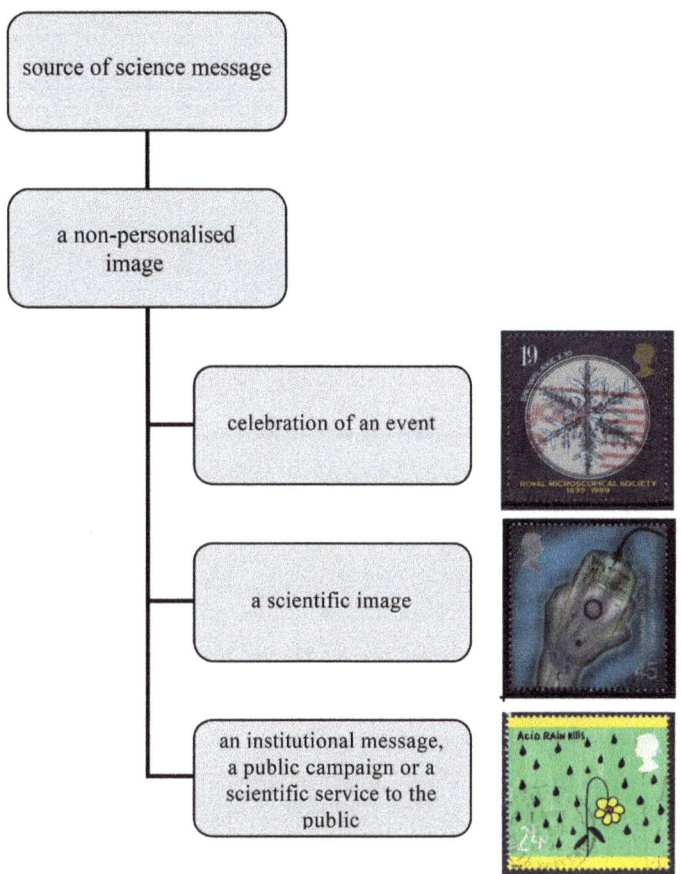

Figure 2.5: Examples of three taxonomy classifications of stamps that celebrate a science.

Source: Author's research.

Celebration of an event: As will be shown below, Great Britain commemorates institutional events and anniversaries of science more frequently than any other country. This set of four celebrates the 150th anniversary of the Royal Microscopical Society (1839–1989). The image examined on the stamp is a snowflake at a magnification of ten times. The other examples in the set show magnifications of different subjects at 6, 300 and 600 times.

A scientific image: A scientific images is used, as will be shown, for one in five of all science stamps. The x-ray image of a hand manipulating a computer mouse comes from the 2001 *Millennium* series, the ninth of 12 projects. This set of four images, titled *Mind and matter*, featured four different designers showing what they had determined as the significant images from four scientific/

research projects. The Millennium Project, containing 48 stamps in all, can be categorised as Royal Mail stamps designed to excite further interest in the observer. The stamp shown is numbered 35 in the Millennium Project series and features the location of Millennium Point, a "science centre in the Digbeth area of Birmingham which aims to encourage youngsters to develop an interest in science and highlights industrial and scientific innovation" (Davies, 2000, p. 60). Scientific images have also been used to look forward into the future, particularly in respect to space fantasies.

An institutional message, a public campaign or a scientific service to the public: This classification, although general, is not a catch-all. It is significant that one in eight science stamps fits here. The example shown, one of a set of four, is from the 1992 *Protection of the environment* issue. This is a noteworthy issue, the first stamp indication from Great Britain highlighting growing concern for the environment. Children's paintings have been selected for the images. Shown is the child's perception of the effect of acid rain. The three other images are the ozone layer, the greenhouse effect and the bird of hope. The objective in issuing the set is obvious as Royal Mail have also described the stamps as *The Green issue*, albeit seven years after Australia Post signalled a growing public concern over the changing climate with its first issue messages on this subject.

A childrens' art competition was the source of images for this set of British stamps looking at the changing climate. The USA and China have also used children's paintings to look into the future, emphasising the fact that a child's perspective is critical if future action is to be taken. Children, or the judges of these competitions, understand the role science will play in solving the problems and effects of global warming. The USA issued a 2001 set of four predictions, entitled *Stampin' the future*. Three of the stamps featured space travel. A Chinese fine arts competition elicited one of four pictures entitled *Enthusiasm for science*.

Context

In conveying its message, the stamp designer has the option of including extra information about the science and the scientist they are featuring in their design. I describe this additional information as it provides context to the image. Context will be studied as a tool towards the telling of the message while, at the same time, providing an entry point for new ideas and concepts. Early stamps showed only an icon of the country of issue, so stamp design has developed in the telling of science messages.

The concept of context, in terms of it providing additional background and enhancing the depth of the message being told on the stamp, is considered throughout this study. There is a strong case to be put forward that context has evolved to the point where it is a major factor of stamp design. The ability to provide meaningful context has improved through technology and the adoption of photogravure printing. Some countries have used context from earlier issues but, to generalise, I believe that the use of context has become more common from the mid-1990s, a date at which internet use became routine and a date that may mark a change in the underlying impetus of science communication philosophy moving from wanting to develop a public understanding of science (PUS), to that of stimulating a public awareness of science (PAS). Context will be discussed from this point forward in my study.

Two examples from East Germany are shown to demonstrate how the addition of context enhances the message. The scientists celebrated are physician and physicist Hermann Ludwig Ferdinand von Helmholtz (1821–1894), and physicist Heinrich Hertz (1857–1894). The simple portrait representations to the left of the figure are from the 1950s. The larger 1994 stamps define the celebrants' life-studies with images of their achievement issued at the time of their death centenaries.

Figure 2.6: Hermann von Helmholtz, *250th anniversary of the Academy of Sciences and death centenary.* **Gibbons catalogue # E24, 2594. Heinrich Hertz,** *Birth centenary and death centenary.* **Gibbons catalogue # E324, 2557.**

Source: Author's collection.

Another example of context is shown in Figure 2.7. This Royal Mail example, entitled *Medical discoveries,* is an extreme example of context in relation to a known scientist. Four acknowledged medical techniques and their proponent scientists are recognised with this issue, but the scientist is not named on the face of the stamp. The celebrants are: *Ultrasonic imaging:* physician Dr Ian Donald (1910–1987); *Scanning electron microscopy,* physicist and electronic engineer Sir Charles Oatley (1904–1996), *Magnetic resonance imaging*, 2003 Nobel Medicine

Prize winner Sir Peter Mansfield (born 1933); and *Computed tomography*, 1979 Nobel Medicine Prize winner Sir Godfrey Hounsfield (1919–2004). The detailed background to each image is provided in the Royal Mail Yearbook (Shackleton, 1994, pp. 36–39). The detail was also provided in the presentation pack for this issue. The image of the achievement has become the main element. This almost changes the classification of the stamp from named scientist and event (achievement) to non-personalised image/scientific image. These stamps use a specialist context to tell their message, more so by Royal Mail than any other postal authority, to date, in telling a medical story.

Figure 2.7: Great Britain, 1994. *Europa, medical discoveries.* **Gibbons catalogue # 1839–1842.**

Source: Author's collection.

From Australia, I am able to reproduce an example of increasing context in association with the scientist's portrait. This is shown in Figure 2.8. From the 1960s, the first single stamp does contain some context. The 1962 *Centenary of John McDouall Stuart's overland crossing,* shows the explorer full face, he is named within the portrait frame, (in very small letters) and the background is very sparse, suggesting an unexplored land. You will note the event is not explained. In contrast, the 2012 two-stamp celebration published 50 years later details the name of the explorer and the event being celebrated, *The overland crossing of 1861–1862*. From the later stamp we also learn that it was a team effort with at least five participants on horseback who raised the Union Jack flag on the shores of the Indian Ocean having travelled from Adelaide. Stuart returned to Adelaide with his team, who actually numbered nine.

2. Developing a Taxonomic Structure

Figure 2.8: An example of the development of context on Australian stamps, 1962–2012. Renniks catalogue # 270. The two stamps on the right are too recent to have been catalogued.

Source: Author's collection.

The designer of the 2012 issue has also been able to incorporate the Stuart expedition into a miniature sheet with the discovery of the Blue Mountains and is able to show the maps of the explorations and date them, thus increasing the context of the issue. The event to prompt the issue is the 150th anniversary of the completion of Stuart's exploration, and, a year later, the 200th anniversary of the *Blue Mountains crossing*, although this is not stated. The issue is entitled *Inland explorers and* is quite explicit in describing the events being celebrated. The four stamps separately show textually more information than the 1960s images, including the event name and date and the names of the participants. None of the available space is wasted, with the image continuing to the very edge of the stamp. The mode of travel is illustrated as is the topography and foliage. The miniature sheet is shown as Figure 2.9.

Figure 2.9: The Australian miniature sheet, *Inland explorers,* which is too recent to have been catalogued.

Source: Author's collection.

Method of analysis within this study

The first step was to list, in chronological order, the country's science stamps, attaching a short narrative of subject matter, and details of whether it was a single stamp or part of a set enveloped within a single theme. Within the description is included a designator, usually an established catalogue number, so that the issue is placed within its context. Within this step was an editing of pseudo-scientific subject matter. For example, is a stamp with just the words "fight against cancer" showing science? Probably not. But if the image includes a microscope and/or a doctor/nurse and/or medical equipment, the representation is beyond a public health announcement and falls within a science framework.

If a scientist is portrayed and named on the stamp, it is determined to have a scientific context, although the scientific achievement might not be described. Can it be assumed that the celebrant is so well known that no additional information need be given? This is examined in the case study titled 'Heroes of Science' in Chapter Six. Listing the scientists honoured on stamps allows for an analysis, by simple count, of which scientists are celebrated by which countries. This may give an indication of how and why scientists were chosen by the postal authority.

Analysis of the messages contained within a postage stamp

The semiotic approach recommended by both Scott (1995) and Child (2008) is based upon the work of Charles Peirce (1839–1914), the founder of American pragmatism, and has been followed in my study. The three elements of the Peirce typology are defined as:

1. Index: a pointer taking the viewer somewhere. An example would be smoke, which is an index to the fire that released it.

2. Icon: a graphic pictorial representation such as a picture, a design, or a photograph. It can be observed for its own aesthetic sake or, more important for our semiotic analysis, analysed to see what the message of the picture is.

3. Symbol: a conventional sign in which elements stand for something else. Thus the symbol "$" stands for dollar, and the post horn is a common symbol for postal service. (Peirce, 1867).

I recognise that in looking at the design of postage stamps these definitions may become blurred and intertwined.

I expected to endorse semiotics as a significant tool to better determine the messages contained within postage stamps issued with a scientific content or context. I have undertaken semiotic analyses of stamps and included a number of these within my study.

Due to the limitations in size, the stamp designer has, perhaps, limited opportunity to show other than the stipulated message.

Understanding the process of stamp development

To help understand the big-picture of policy for a postal authority, I have reviewed the process of stamp development through documents taken directly from the UK Royal Mail and the United States Postal Service websites. I have also taken advantage of the opportunity to look through the archival corporate files of Australia Post and Royal Mail, which included the documentation of issues deemed to be scientific. I also interviewed the curators of these archives.

While undertaking my study, I was fortunate enough to be asked by (the UK's), *Royal Mail Philatelic Bulletin* to write an introductory article for a 2010 stamp issue of medical breakthroughs. To facilitate this process, I was sent the initial designs and concepts for the issue and was sworn to secrecy about the particular scientists and breakthroughs that were to be celebrated. The set is reproduced in the discussion in Chapter Eight, Figure 8.1.

Interviews with significant contributors to my study

I had approval from the ANU Ethics Committee to interview the heads of the stamp units of Australia Post, New Zealand Post and Royal Mail for my research. The encouragement and advice from these managers was open and frank, within the commercial constraints that were explained to me. All three of the postal administration managers were interested in my study and endorsed my research interest as appropriate and have requested sight of my eventual findings. I was given copies of the latest stamps that each authority determined as related to science, when I met them. These were a valuable aid in our discussions. I raised the flora and fauna issue, described in the next chapter, with each manager and their response was an enthusiastic, 'Yes, that's science', if the stamp contained the image's botanical name as text. My pursuit of them to tell me the actual science message confirmed to me, with their eventual agreement, that many flora and fauna stamps are issued to attract the casual buyer and the thematic philatelist within a tourist attraction and advertisement classification. The managers confirmed that each authority has, as an objective, the sustainability and a continuing profit for the stamps unit.

I have also been privileged to meet and discuss my study with two of the three authors of the seminal books that were essential reading, Professors Altman and Scott. I exchanged notes with the third, Professor Child of the American University in Washington, but he was obliged to cancel our scheduled meeting due to health problems.

Semiotics has provided the prompt in opening discussions with all interested parties as I pursued my study. I used a reproduction of the 1980 *Second National Conference of the Chinese Science and Technology Association* stamp issue—already shown in the opening chapter and replicated and shown in Figure 2.10—on all my ANU Ethics and third party communications. It has remained of importance through my project. On the surface it is a design that shows apsaras and symbols of modernisation.

Figure 2.10: China, 1980. *Second National Conference of the Chinese Science and Technology Association*. Gibbons catalogue # 2974.

Source: Author's collection.

The question is, what are apsaras—traditional Chinese mythical wood-nymphs—doing on a Chinese stamp just four years after the Cultural Revolution, when the Red Guards reviled and rejected everything intellectual? The stamp image is full of symbols of modern technology and the main figure is pointing towards the year 2000 which is spelled out and underlined by the space rocket. Stars are included in the background, as are the symbols of atomic structures, a gear-wheel and a symbol of the Association being celebrated. The text designating the indices of the stamp, country and service value and the purpose for the issue are unobtrusive, but are available if someone is seeking the information. My showing and reading of this stamp has been a conduit into discussion with all the people I have interviewed for this study.

Not many stamps convey a message other than the obvious, but a semiotic analysis is appropriate for some. I also carried, as back up to further prompt to discussion if necessary, reproductions of the three Polish stamps shown in Figures 2.11, 2.12 and 2.13.

Figure 2.11 celebrates the achievement of Ignacy Lukasiewicz. The shape of his petrol lamp is shown in white as the background to his image. This scientist has been the subject of three Polish stamp issues. His life dates are shown in text. The stamp also features a congress being held in Poznań, Poland during 1960, although no overt link is made between Lukasiewicz and the pharmaceutical industry.

Figure 2.11: Poland, 1960. Ignacy Lukasiewicz (1822–1882), inventor of the petrol lamp and the *5th Pharmaceutical Science Congress in Posnań*. Gibbons catalogue # 1172.

Source: Author's collection.

Figure 2.12 shows an arrow piercing an "E". The text describes the celebration of the *50th anniversary of the deciphering of the German Enigma cipher codes*. No explanation of who or what Enigma is has been made. It has only been later in time, during the 1980s, that the full story of Poland's contribution to the deciphering of German Military codes during WWII at Bletchley Park in the UK has been made public. The colours of the arrow are the same as that used in the Polish flag. In 2009, a second stamp, part of a celebration set of *Poles in the World* names three of the Polish scientists and includes photographs. The Enigma machine mathematician/cryptographer code-breakers recognised are Jerzy Witold Różycki (1909–1942), Marian Adam Rejewski (1905–1980), and Henryk Zygalski (1908–1978).

Figure 2.12: Poland, 1983. *50th anniversary of the deciphering of "Enigma" machine codes*, and Poland, 2011. *Polish scientists in the world*. Gibbons catalogue # 2889 and WNS catalogue # PL039.09.

Source: Author's collection.

Figure 2.13 needs no elaboration. It uses an iconic image of Marie Curie at work in her laboratory. It is an image that has also been used by France and Ireland. It is unencumbered except for the two required indices of country and value. The colours are subtle and subdued.

This stamp is actually one of a set of two and was also published as a miniature sheet titled *Centenary of the Chemistry Nobel Prize for Marie Curie Skłodowska*. The other, smaller stamp is equally simple, featuring three images: what Marie Curie is seeing in her test tube, overlaid by a gold Nobel Medal, and the profile of Alfred Nobel.

Figure 2.13: Poland, 2011. *Centenary of the Chemistry Nobel Prize for Marie Curie Skłodowska*. WNS catalogue # PL036.11.

Source: Author's collection.

During my interview with him, the Manager of Stamp Strategy of Royal Mail introduced me to his concept of the communication message on a stamp being a mirror or a lens. My definitions of these terms are described later. I felt obliged to consider using this additional classification in my formal taxonomy. At the time, however, I did not include this classification because I believed that it was too subjective. I have subsequently undertaken surveys to ask colleagues and members of the public to consider my definitions and judge where 16 science stamp examples would be placed a on scale with mirror at one end and lens at the other. The results showed very little consistency of response. The date at which the stamp had been issued was proved to be the most significant. A stamp foretelling a future event—the image of first moon rocket, for example—is seen as a lens at the time, and as a mirror of an achievement after the event.

Access to specific (stamp issue) Post Office development files

Australia Post

Prior to my second meeting with Australia Post historian and curator Richard Breckon in Melbourne in October 2010, I requested access to the files of four stamp issues from the Australia Post archives. As it happened, the curator was able to provide only two of the requested four files. partly because of the 30-year embargo on public service files. Knowing this, Mr Breckon had prepared a précis of file contents of the two other issues for me.

Royal Mail, UK

I also had recourse to Great Britain files through the British Philatelic Museum and Archive (BPMA) during May 2011. Prior to my attendance at BPMA, curator Douglas Muir had made available specific academic studies of the issue of stamps I have designated scientific. The procedures seem to be straight-forward and in accord with the published process.

Libraries and postal museums

During the course of my study, I have been able to inspect specialist philatelic items contained within the National Library of Australia, The British Museum, the Irish National Philatelic Museum, and the Library of Congress and the Smithsonian Library and Museum in Washington to facilitate my understanding of my subject.

During the development of this methodology, I did consider whether I should study a control group against the science taxonomy. The obvious contrast would have been to compare art stamps. This decision would have been even more subjective. I decided that the count within ten different countries would provide the control.

3. Classification and Analysis

> Designs in connection with postage stamps and coinage may be described,
> I think, as the silent ambassadors of national taste (W. B. Yeats).

In this chapter I shall follow the sequence of the methodology in outlining the general results from my study. I have examined the stamps of the ten countries of interest and determined which have a science message, defined by examining each image to ask: Does the stamp name or recognise a particular scientist as the main or significant image of the design? Is the subject image non-personalised, showing a science, a scientific device or the description of science in the service of the public to convey its message?

Following the determination that these stamps were showing a science as its message, through portrayal of a scientist or by a science in the abstract, I classified the stamps according to my taxonomy and reviewed the characteristics of the stamps of each country.

Before investigating the results in detail, I shall use selected images to further flesh out the steps described in the methodology. Figure 3.1 expands upon the description of the use of a scientific artifact to confirm that the stamp contains a scientific message. This message is possibly ongoing, as the microscope and stethoscope images support an underlying public health and awareness campaign.

Figure 3.1: United States, 1965. *Crusade against cancer*, a single stamp. Scott Catalogue # 1263.
Source: Author's collection.

If a scientist is portrayed and named on the stamp, it is determined to have a scientific context, although the scientific achievement might not be described. Can it be assumed that the image of the celebrant is so well known that no additional information need be given? I address this question in the 'Heroes of Science' case study in Chapter Six. As an example, Einstein is the scientist most often shown on postage stamps from all the countries of the world. He was, indeed, the face chosen to represent the 2005 International Year of

Physics. Einstein was German-born and much of his work was conducted in Germany before he went to the US. We can deduce that the US is claiming Einstein as an honorary American through showing him on their definitive (everyday) stamps after taking US citizenship in 1940, at the age of 61.

Figure 3.2: United States, 1965. *Albert Einstein (1879–1955), physicist.* **Scott catalogue # 1285.**

Source: Author's collection.

A single scientist may be seen in a different guise by the designer, whose approach will have been defined by the issue intent and message to be illustrated. For example, the US has celebrated the life and work of French/American ornithologist John James Audubon (1785–1851) three times: as one of five celebrities in the 1940 *American scientists* issue; on a 1963 single stamp; and as one of 26 scientists described as the *Great Americans* set of 1980–1985.

Figure 3.3 illustrates how the stamp designer is able to use the different formats available to the issuing authority to represent the scientist or their achievement. The first stamp is from the 1940 set of *American scientists* which was issued at the beginning of WWII, confirming the place of the US in a difficult world. There is no overt reason for selecting the Audubon painting for issue in 1963, or the image of Audubon as part of the definitive issue of 1980. We do know, however, that events and anniversaries are a common prompt for an issue.

Figure 3.3: United States, 1940, 1963 and 1980. *John James Audubon.* **Scott Catalogue # 874, 1241 and 1863.**

Source: Author's collection.

Medicine and public health issues

I have never lacked belief that medicine research and practice, and veterinary practice fall within the definition of science. Public health issues have required some thought. The postal authorities have issued some very graphic warnings against smoking and drug taking. Under the umbrella of my study, the representation of science, I ask: are these messages scientific? In each, admittedly subjective, evaluation, I have asked this question. I have also tried to determine if the message shows science in support of a social issue. With very few exceptions, I have accepted public health messages as scientific and seeking public well-being. Climate change awareness is another social issue that will be pursued as a facet of science awareness in this book.

Another issue: are flora and fauna images on stamps sending a science message? I was obliged mid-study to challenge my classification of science. Early on, I made the decision that simple representations of flora and fauna on stamps would not be deemed science. The question was raised as to whether inclusion of the botanical, scientific name of flora and fauna should define stamps as scientific. My initial decision was "no", the fact that Latin is unfamiliar to many of the general public does not contribute to an awareness of science. Put in other words, a warm, fuzzy or colourful representation of flora and fauna is not made scientific simply because they are named. I remain convinced that these regular issues do not fulfill my definition of science. Sets are issued every year under many guises, including all sorts of pets, cats, dogs, working dogs, race and working horses, tropical fish and birds. I also believe that the regular issue of stamps showing dinosaurs are not promoting science, but are issued to attract the thematic collector and children. This evaluation worked quite well until it was challenged through an insight given to me by Philip Parker, Head of Stamp Strategy at Royal Mail in London. Mr Parker was telling me about the public reaction to the set of stamps shown in Figure 3.4. He believed that the use of the taxonomic name on this issue had endowed the stamp with the epithet of scientific, and agreed with my assessment that the impetus for the issue was to issue a strictly scientific message. It showed the decline in numbers of the subject species quite dramatically though graphs as the background to the image. The set was the subject of Royal Mail market research, post issue, to which I was given access.

The Representation of Science and Scientists on Postage Stamps

Figure 3.4: Great Britain, 1998. *Endangered species*. Gibbons catalogue # 2012–2020.

Source: Author's collection.

The *Endangered species* set has the text "Decline in distribution" at the top of every stamp and incorporates a graph representing the decline in the numbers of these species in the background. I had actually included this set in my initial evaluation and in the Great Britain count for my study because I had thought it scientific in context, not because the botanic names were shown, but because of the *Endangered species* connection. Mr Parker explained that the set had been very unpopular and that market research had indicated that the buying public was not keen to buy stamps that conveyed negative news. Paradoxically, the 20p (dormouse) stamp was voted the most popular stamp of 1999 by readers of Royal Mail's *Philatelic Bulletin* and the 31p (song thrush) as number two. But the readership is a small and specialised sample group. As a direct result of the negative response in the market research, Royal Mail subsequently engaged upon a sustained series of sets of ten stamps to tell an affirmative message showing how species are increasing as a result of positive action. I would not have included the set shown in Figure 3.5 in my count without the insight of Mr Parker's comments. It is now included, as are the other subsequent four issues in the series that refute the message given in the 1998 set. He went further, to question whether the good news may have perhaps been exaggerated on the later issues. The red kite NVI first class stamp, shown in the middle of the top row of Figure 3.5, celebrates the fact that five red kites have been re-introduced into Wales, but it is not known how successful this reintroduction has been. The ten stamps in Figure 3.5 show the bird being featured with text giving the number of the species from two given dates: an early census and a later count.

The subsequent four issues in the Royal Mail *Action for species* series have shown the following themes within the ten different stamps that are included as science stamps:

- 2008, second series: *Insects*
- 2009, third series: *Plants* (the ten stamps in the set also celebrate the 350th anniversary of the Kew Royal Botanic Gardens)
- 2010, fourth series: *Mammals*
- 2011, fifth series: *World Wildlife Fund (WWF), protecting the natural world*.

Figure 3.5: Great Britain, 2007. *Action for species (first series): birds*. Gibbons catalogue # 2765–2773.
Source: Author's collection.

The flora and fauna discussion: Further input

Ivor Masters of New Zealand Post had asked his staff to define science on New Zealand stamps prior to my meeting with him during March 2012. He told me that they had decided that use of a textual botanic name signifies a scientific intention for the issue. They had, in fact, provided him with stamp issues from the previous 12 months to make their case. We looked through these examples and agreed that it was a grey area. He acknowledged that there were several themes on stamps that were regularly issued with the thematically inclined philatelic customer in mind. Flora and fauna also fell into his category of tourism promotion. We also looked at a set of five dinosaur stamps, entitled *Ancient reptiles of New Zealand*, that had been issued in 2010. The stamps are large and were also sold in a miniature sheet that gave supporting information to the image. One of these stamps is shown below in Figure 3.6. Included in the figure is a similar stamp from 2009, again from a set of five, called *Giants of New Zealand*. We agreed that despite showing fauna and including a taxonomic name, the 2010 issue did not send particular science message. The message was principally educational, explaining that the animals once inhabited New Zealand.

Figure 3.6: New Zealand, 2009 and 2010. Single stamps from two sets: *Giants of New Zealand* and *Ancient reptiles of New Zealand*. Campbell Paterson catalogue # S1132 and S1181.

Source: Author's collection.

Having thought it through, I retained, in essence, my original decision: flora and fauna will not be counted unless their relationship to a scientific message is clear. An exception to this rule is made when the celebration of a named scientist includes flora or fauna. One such example is shown as Figure 3.7. This miniature sheet of four different images of the work of John J. Audubon, issued by the French Post Office in 1995, clearly honours the work of the ornithologist and artist. Audubon is also shown on three separate issues of the United States Postal Service as previously shown as Figure 3.3. One can only wonder if France is laying claim to Audubon as a French national with this issue. Audubon was born in the French colony of Saint-Dominique, now Haiti, and lived in France until emigrating to America at age 18.

3. Classification and Analysis

Figure 3.7: France 1995, *John James Audubon (1875–1851), American ornithologist*. Gibbons catalogue # MS3253.
Source: Author's collection.

Figures 3.8 and 3.9 offer examples of where flora falls within the classification of supporting a scientific message. Figure 3.8 is a set of three Russian stamps from 1956, *Ivan Michurin, biologist, birth centenary*, which show flora in the context for the celebration. Figure 3.9 reproduces two stamps issued by the United States in 1997 to show the scientific prints of botanist Anna Merian (1647–1717), the German-born entomologist and scientific illustrator. The Russian stamps show flora to place Michurin's achievements in context, as well as showing his life-dates and the reason for the timing of the issue. The US stamps show flora as scientific examples of the species, and although neither Anna Merian nor the species are named, the message is scientific. Merian's work has been accepted as a standard of excellence in botanical illustration and has been used in plant taxonomy, classification and identification since the seventeenth century.

Figure 3.8: Russia, 1956. *Ivan Michurin, biologist, birth centenary*. Gibbons catalogue # 1968–1970.
Source: Author's collection.

Figure 3.9: Anna Merian, 1997. *Merian botanical prints*. Scott catalogue # 3126–3127.
Source: Author's collection.

In summary, therefore, my policy is to exclude from my study any flora and fauna stamp that is not part of the evolving story of science. I shall not consider stamps that, for example, show working dogs, domestic animals or butterflies, although many countries regularly issue such subjects. The reproduction of any image used to enhance the awareness or understanding of science is included.

Should I classify explorers as scientists?

Generally speaking, the postal authority will celebrate the achievements of its native heroes. However, there is a potential area of overlap between local and foreign heroes. Poland's local hero may well be a foreign hero to Australia, celebrated by both countries. One explorer who meets this description is Sir Edmund de Strzelecki (1797–1873), scientist and explorer of Tasmania, who was celebrated by Poland in 1973 and by Australia in 1983. Many explorers have such a dual identity. Exploration is a vital component of a country's history and all postal authorities have featured explorers as examples in their nation-building messages. Explorers were important people in a world of opportunity during the eighteenth and nineteenth centuries, when new lands were being opened up. Men of stature and enterprise were encouraged into this role. But were explorers scientists? In his introduction to the Encyclopedia of World Explorers, Waldman writes of European countries' political and financial motivations behind exploration: "The country that financed such expeditions wanted a good return for its money. It took personalities of integrity such as Alexander von Humboldt (1769–1859) … to cultivate pure scientific curiosity" (Waldmann, 2002). I accept Waldmann's comments and have decided to include earlier explorers on the basis that they were the polymaths of their time whose work was dependent on the latest science and technology. McCalman (2009) writes of how four voyagers to Australasia won the battle for evolution and changed the world: Charles Darwin (1831–1836), Joseph Hooker (1839–1843),

Thomas Huxley (1846–1850), and Alfred Wallace (1848–1866). European explorers are extensively celebrated by Australia, New Zealand and the United States as well as those countries originally sponsoring the explorers.

I have, therefore, included explorers as scientists on the basis that much of the exploration of the nineteenth and early-twentieth century had a scientific incentive in addition to commercial ones. Exploration implies a lack of knowledge of what might be found and that it is conducted by countries with a political, as well as a scientific motive. Another viewpoint is contributed by Harding (2011):

> Moreover, Steve Harris argues that historians have been preoccupied with the wrong sciences when considering how Europe was able to advance so rapidly. It is these "long distance sciences" — that is, the explorations and enquiries that were part of the voyages themselves that created important kinds of changes in knowledge-seeking practices to which we remain indebted today, but which have been neglected by mainstream histories of science … The voyagers and their sponsors needed astronomies of the southern hemisphere, better oceanography, climatology, marine engineering, cartography. And botanical, medical and ethnographical knowledge (Harding, 2011, p. 35).

The count of the total number of the stamps of ten countries

In order to provide a context for my later analysis of science stamps I have gathered data on the total number of stamps issued by the countries of interest.

The different countries of the world have different dates for the introduction of postage stamps. My study starts with the first stamps issued and continues through to the end of 2011 in order to provide a consistent cut-off date. For the countries whose stamps are not yet catalogued to that date, I have estimated a final count assuming that the number of stamps issued will follow the pattern of previous years. I have included the images of stamps from 2012 and 2013, where I have been able to source them, in my text, but they do not appear in the number count.

A simple analysis of the number of stamps issued in a year by the countries is shown in Table 3.1.

Table 3.1: The total number of stamps issued by all nominated countries from earliest issues to 2011.

Country	Time-span	No. of years	No. of stamps issued	Average no. of stamps per year
Australia	1901–2011	111	3,338	30
New Zealand	1855–2011	157	2,473	16
Great Britain	1841–2011	171	3,051	18
France	1849–2011	163	4,197	26
Germany to 1945	1871–1945	75	898	12
East Germany	1949–1990	42	2,852	68
West Germany	1949–1990	42	1,081	26
Reunified Germany	1991–2011	21	1,408	67
Ireland	1922–2007	86	1,763	21
Poland	1860–2011	152	4,086	27
Russia	1857–2011	155	7,055	48
People's Republic of China	1949–2011	62	3,990	64
United States of America	1847–2007	161	4,219	26
Australian Antarctic Territory	1957–2011	55	106	3
NZ Ross Dependency	1957–2011	55	127	2
British Antarctic Territory	1963–2011	49	581	12
French Antarctic Territory	1955–2011	57	725	12
Total			41,950	

Source: Author's research.

For consistency, this table is derived from the latest Gibbons and Scott Catalogues, and from the UPU WNS system where possible for those issues not yet published in a Gibbons or Scott catalogue. For issues not recorded or countries that do not subscribe to the WNS system, data is taken from Stanley Gibbons' Monthly catalogue supplements. The numbers are simplistic as the total number shown is that of the issues available from the local post office in the subject country. Some authorities also issue what they call special airmail stamps, parcel-post stamps and semi-postal stamps. These have been excluded, although image examples are used in the study where appropriate.

I have included the Antarctic Territory stamps issued by Australia, New Zealand, Britain and France on the basis that polar research in the Antarctic is a significant feature of these issuing authorities' stamp issues. The French authority embraces the Antarctic Adélie Land and four additional South Indian Ocean possessions. I did consider Antarctica as a separate case study but as the results are fairly consistent across the four authorities I have included the results within this chapter. The unique political messages being sent are discussed in Chapter Four.

It would appear that several countries have over time issued a higher average number of stamps each year. New countries issue stamps to legitimise and publicise change and to establish identity. These include East Germany (1949–1990), which issued stamps at a time when all countries were increasing stamp frequency, following the trend of using stamps as message communicators. East Germany also pursued the celebration of issues featuring Russian space research during this time. Post-1991 reunified Germany has, over the past 20 years, issued an average of 67 new stamps each year. The People's Republic of China is another newer country with a high frequency of issues. Mackay comments upon the number of stamps issued by Russia: "Under the Soviet regime more than 6,000 stamps were issued, but in the years since the collapse of communism a more moderate policy has been pursued, omitting the political propaganda of the former era" (Mackay, 2011, p. 153).

The average number of Australian stamps issued is somewhat exaggerated by the issue of stamps of gold medal winners from such sporting events as the Olympic and Commonwealth Games. To celebrate the 2006 Commonwealth Games, held in Melbourne, for example, 122 commemorative stamps were issued, inflating the overall averages.

As suggested above, the number of stamps issued in recent years has been higher than in early years. This policy change is not readily apparent from the reading of this table, so a separate count has been made to calculate the average number of stamps issued each year during 2002–2011. Table 3.2 illustrates how the countries with the longest issuing history—Great Britain, France and the United States—have significantly increased their revenue potential, (within an environment where the use of traditional postal services is declining), especially from the collector (philatelic) market. It is worth noting that Russia's increase is more modest, but is coming off a high base, and that the Russian Post Office has followed a policy of using personal images over its history certainly longer than Great Britain and the United States, whose issuing policy places limitations upon named individuals who can appear on the postage stamp. I shall discuss how technical changes might make a difference to the ease of stamp production later in this book. I do not perceive the technology of stamp production as having a direct influence upon the number of different stamps issued. Despite the inroads that the internet, email and social networks are making into overall mail deliveries, the number of stamps being issued by all countries continues to rise as a result of the new market model to optimise postal services sales into the leisure and casual buyer markets.

Table 3.2: The increase in total number of stamps issued in the twenty-first century, compared to the 1800s and 1900s.

Country	Average number of stamps issued Each year		
	Since first issue	2002-2011	Increase
Australia	30	110	3.6 x
New Zealand	16	84	5.2 x
Great Britain	18	113	6.3 x
France	26	150	5.8 x
Germany to 1945	12	N/A	N/A
East Germany	68	N/A	N/A
West Germany	26	N/A	N/A
Germany Re-unified	67	68	0
Ireland	21	63	3 x
Poland	27	60	2.2 x
Russia	48	85	1.8 x
People's Republic of China	64	94	1.5 x
United States of America	26	134	5.1 x
Australian Antarctic Territory	3	5	1.6 x
NZ Ross Dependency	2	5	1.6 x
British Antarctic Territory	12	26	2.2 x
French Antarctic Territory	12	25	2.1 x

Source: Author's research.

Taxonomy results

The first classification of the taxonomy for my study was to look at the science stamp and note whether it was a unique stand-alone conveyor of the message or part of a set. The results appear in Table 3.3. Also included within this table is the record of the number of colours used in the printing. Technical advances since 1840 have facilitated the use of multiple colours in printing, essentially from 1883. From this point in my study, tables will refer to the numbers of science stamps.

3. Classification and Analysis

Table 3.3: The percentage of all science stamps issued as a single stamp or part of a set and the number of colours used in their printing.

Country	Single stamp	Set or part of a set	Single-colour image	More than one colour used
Australia	25.5%	74.5%	15.7%	84.3%
New Zealand	18.0%	82.0%	4.5%	95.5%
Great Britain	2.5%	97.5%	0.4%	96.6%
France	47.2%	52.8%	17.9%	82.1%
Germany	38.0%	72.0%	17.7%	82.3%
Ireland	7.1%	92.9%	21.4%	78.6%
Poland	20.2%	79.8%	14.2%	85.8%
Russia	31.4%	68.6%	14.3%	85.7%
China	12.8%	87.2%	10.7%	89.3%
USA	29.7%	70.3%	24.8%	75.2%
Overall average	27.4%	72.6%	14.6%	85.4%

Source: Author's research.

Figure 3.10 represents the proportions, country by country, of those stamps issued as a single or as a part of a set. The data are from columns two and three of Table 3.3.

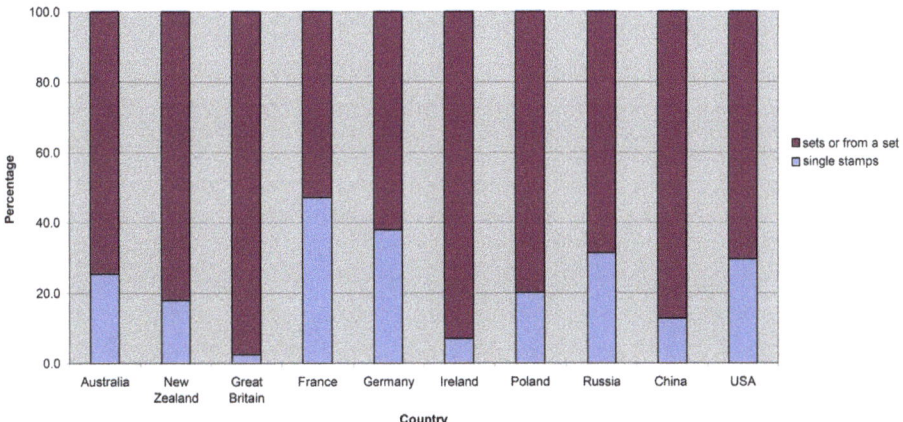

Figure 3.10: The proportion of science stamps issued as singles or as part of a set, by country.
Source: Author's research.

Whether the stamp is issued as a single or as a part of a set is determined by the issuing authority's policy as well as the designer's recommendations. The single is by definition a one-off. The scientist whose image is portrayed is named and sometimes their life-dates are shown. France—followed by Germany, Russia and the United States—issues more single science stamps than any other country. Great Britain issues the fewest single stamps, reflecting a policy through which it issues the pre-payment tokens at the most commonly purchased standard postal rates as a set. Great Britain also tends to celebrate the anniversaries of institutions, including named scientists who have contributed to that institution's reputation. European countries, Great Britain, France, Germany and Russia have longer scientific histories and therefore a greater opportunity for anniversary celebrations. In theory, the value of a set of stamps should allow for the development of a story across a number of images but the opportunity for a member of the general public to view the set has to be limited, not so the collector. Thus the designer will have in mind the need for each stamp of the set to stand alone, as seen on an envelope, and convey at least part of the overall message.

I have not created a chart to show differences in the colours used on a country comparison basis as they appear to be consistent and related to time and technology. Science on stamps occurred when multicolour printing was the norm. However, the stamp designer will sometimes use a single colour in order to indicate that the stamp has a heritage connotation. I cannot show that science stamps are more or less colourful than other stamps issued, except perhaps when looking at the space research stamps of the Eastern Bloc from the 1960s to the 1980s, which are certainly very colourful (see Figure 4.43 as a typical example). An aluminium stamp was issued in 1961 (Figure 4.27) with the title of *Cosmic flights,* which was certainly quite different.

I believe it is worth noting an evaluation of the most expensive stamp set produced by Royal Mail, noted by its Head of Stamp Strategy during my discussion with him in May 2011. The issue was not well received by the public. This is an example of production being integral to the presentation and a part of the story being told. The set concerned is the Great Britain 2001 set of six stamps celebrating Nobel prizes. The three science prize representations are shown in Figure 3.11. The stamp representing chemistry is a heat sensitive image of the carbon 60 molecule, celebrating the work of Sir Harold Kroto (born 1939), Nobel Prize winner in 1996. The economic sciences example emphasises the texture of intaglio printing. The peace image was an embossed dove carrying a maple leaf. The physiology or medicine image releases a eucalyptus scent when scratched. The literature stamp requires the use of a magnifying glass to read the poetic text. The physics representation is a hologram of electrons orbiting in a boron molecule to celebrate Dennis Gabor (1900–1979), the Hungarian-born inventor of holography, who was a Nobel laureate in 1971. The set was not popular. Was it too hard for the public to understand the effort that had been taken to illustrate the science?

3. Classification and Analysis

Figure 3.11: Great Britain, 2001, Three science *Centenary of Nobel Prizes* stamps, from the set of six, representing chemistry, medicine and physics. Gibbons catalogue # 2232, 2235 and 2237.
Source: Author's collection.

The representation of science on postage stamps

Table 3.4 shows the total number of stamps issued and the number classified as illustrating science and, therefore, sending a science message. The third data column shows the percentage of science stamps issued of the total issued. As previously mentioned, I have used the cut-off date of the end of 2011. This is the last year in which I have data confidence as published catalogues and even the postal authority lists can be delayed.

Table 3.4: Total numbers of postage stamps issued and number of science stamps, by country.

Country	Total stamps issued (to 2011)	Total science stamps (to 2011)	% of science stamps
Australia	3,338	222	6.4%
New Zealand	2,473	103	4.2%
Great Britain	3,051	318	10.4%
France	4,197	516	12.3%
Germany	6,239	552	8.8%
Ireland	1,763	91	5.2%
Poland	4,086	272	6.7%
Russia	7,055	1,305	18.5%
China	3,990	337	8.4%
United States	4,219	308	7.3%
Total	**40,411**	**4,024**	**9.96%**
The Antarctic Territories of			
Australia	106	71	67.0%
New Zealand	127	47	37.0%
Great Britain	581	333	57.3%
France	725	310	42.8%
Total	**1,539**	**761**	**49.5%**

Source: Author's research.

We might deduce that one reason that the overall higher percentage of the older European countries is high is their science legacy. As the early intellectual powerhouses of scientific endeavour, they have the celebrities and achievements to publicise at anniversaries of scientists births and deaths, and anniversary of scientific events and achievements. A major influence on the average of one stamp in ten representing science is the high average of Russian stamps and the high number of stamps Russia issues. The Soviet policy of stating public ambitions on stamps is discussed in the next chapter.

The very high proportion of the stamps reflecting science on the stamps of Antarctica is examined later in my study, as is the political motivation of emphasising scientific research as a justification to assume sovereignty of the territory.

The overall number of stamps illustrating a science has increased over time, keeping pace with the total number of issues, as is shown in Table 3.5. There is a trend towards a higher proportion of all stamps in the modern era representing science. The two exceptions to this are Germany (all constitutions) and the United States. The former's later figures reflect the reunification of East and West Germanies and changed political objectives. The former East Germany had been very much influenced by Russian policy and issued many space research issues, as shown in Table 3.6. The reason for fewer United States stamps showing science is not so easily diagnosed, but is influenced by the United States Postal Service policy of not celebrating living persons and of waiting for several years before celebrating any individual. These influences are discussed later.

Table 3.5: Table showing the proportion of issued postage stamps representing science over three time periods.

Country	% of science stamps each year		
	Over many years	2002–2011	2007–2011
Australia	4.6%	4.1%	6.9%
New Zealand	3.6%	3.0%	6.9%
Great Britain	9.0%	10.6%	16.7%
France	5.1%	8.3%	10.8%
Germany (all constitutions)[1]	8.8%	8.7%	7.2%
Ireland	3.9%	4.6%	7.8%
Poland[1]	6.2%	6.5%	10.0%
Russia[1,2]	15.5%	14.7%	16.6%
People's Republic of China	7.2%	8.4%	9.8%
United States of America[3]	7.2%	2.8%	2.5%

Notes: 1) The number of science related stamps from the Eastern Bloc countries are inflated by the space-race issues of the late-1950s to the 1980s, as reflected in Table 3.6.

2) Soviet Russia was broken up in 1991. The figures in the 2002–2011 period show stamp numbers from the smaller Russian Federation, which is currently issuing about half the stamps that it did as the USSR.
3) I am unable to state why the percentage of science stamps has declined for the United States. I perceive a distinct move for the United States Postal Services to self-adhesive stamps, which tend to be smaller and, therefore, allowing less scope for context. In this study, I discuss the use of context, through which stamp designers follow the understanding and awareness of science by the general public. Looking through all of the stamps issued by the United States Postal Services from 2000, there is a trend towards messages celebrating the United States through its own emblems and institutions.
Source: Author's research.

Table 3.6: Table showing the proportion of science stamps illustrating space research over the life of postage stamps.

Country	No. of science stamps	% of space research themed stamps
East Germany	260	24.6%
Poland	253	26.5%
Russia	1,096	30.0%

Source: Author's research.

The combined taxonomy results

Figure 3.12 shows the overall results of classifying science stamps from the ten countries being studied. Each country has its own profile, which reflects the issuing authority's policy. Great Britain, for example, issues very few stamps as singles. One can generalise to state that Great Britain will issue ten sets each year, one of which will most likely have a scientific theme, and the set will comprise the four or five basic, most sold values, representing its current scale of services and charges.

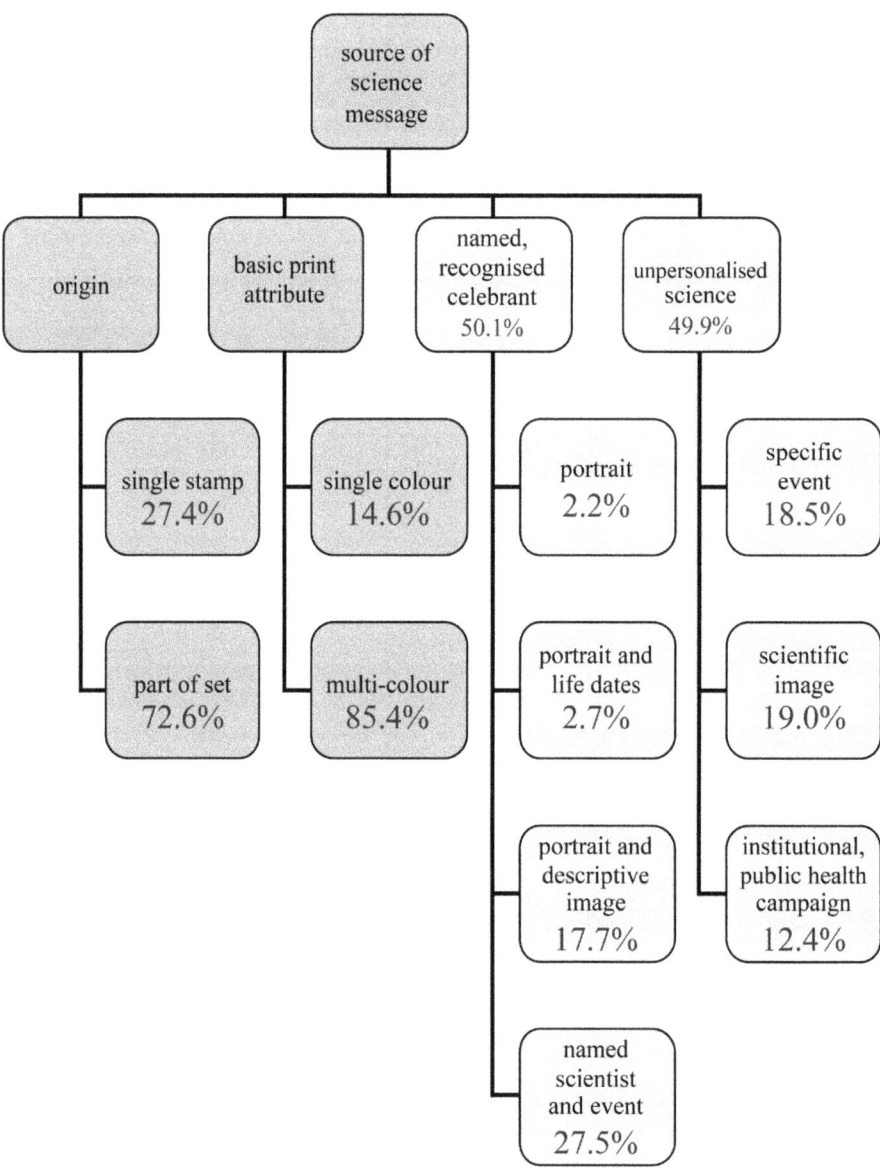

Figure 3.12: A basic taxonomy of the stamps in this study, derived from the conceptual representations of Kress and Van Leeuwen (2006).
Source: Author's research.

Stamps representing a named and recognised scientist: A human dimension

Table 3.7 shows the percentage of all science stamps issued featuring a scientist, 50.1% of all the science stamps. I have not included Antarctic Territories in this table, as they are discussed as a separate entity elsewhere.

Table 3.7: The percentage of all science stamps issued featuring a scientist.

Country	Portrait	Portrait and life-dates	Portrait and descriptive image	Named scientist and an event
Australia	3.3%		35.9%	17.6%
New Zealand			33.7%	11.2%
Great Britain	0.7%		31.3%	30.9%
France	0.5%	13.7%	32.5%	20.3%
Germany	3.7%	3.2%	4.6%	38.2%
Ireland	2.9%		50.0%	2.9%
Poland	3.2%	11.1%	11.5%	25.3%
Russia	0.3%	1.2%	12.6%	34.8%
China			16.6%	8.3%
USA	10.1%	0.3%	23.2%	18.6%
Average	2.2%	2.7%	17.7%	27.5%

Source: Author's research.

Science stamps incorporating a named scientist and an event is the most popular classification of all the science stamps in the study. Almost 40% of all German stamps are issued in this format, followed by Russia (34.8%) and Great Britain (30.9%). Countries with a lower than average representation in this classification—Ireland, New Zealand and perhaps Australia—have fewer local scientists of repute whose anniversaries might be celebrated. China generally uses fewer messages with a human dimension and prefers science in abstract. It would appear that the United States prefers images that include portraits and some description of the scientists' achievement rather than anniversaries and events.

Stamps representing a science rather than a scientist: Science in the abstract

Table 3.8 shows the percentages of stamps issued featuring a science, 49.9% of all science stamps.

Table 3.8: **The percentages of stamps issued featuring a science.**

Country	Event/anniversary	Scientific images	Institutional message
Australia	13.1%	22.2%	7.8%
New Zealand	28.1%	21.3%	5.6%
Great Britain	0.7%	8.4%	28.0%
France	7.1%	12.7%	13.2%
Germany	28.7%	15.2%	6.5%
Ireland	8.6%	11.4%	24.3.0%
Poland	11.5%	23,3%	14.2%
Russia	22.7%	20.5%	7.9%
China	24.5%	19.7%	31.0%
USA	10.8%	30.1%	6.9%
Average	18.5%	19.0%	12.4%

Source: Author's research.

Figure 3.13 combines the overall results shown in Tables 3.7 and 3.8 to illustrate the proportions of each country's images that feature a named and recognised scientist or show science in the abstract.

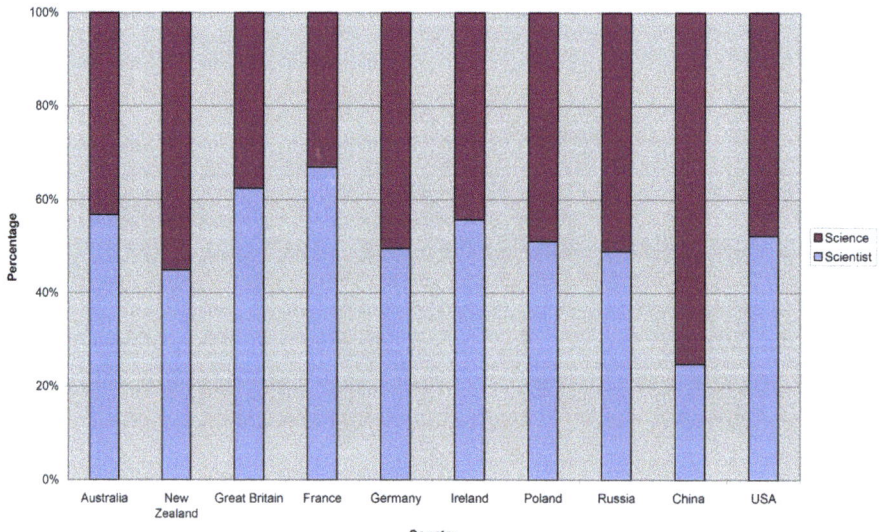

Figure 3.13: The focus of the image/message on the science stamps, by country.

Source: Author's research.

Approximately half (50.1%) of the science stamps issued include a named, recognisable scientist within their images. The other images show a science in abstract (49.9%) and relate to a specific event, include a scientific image, or contain a message that is institutional, related to a specific public campaign or show science as a service to the general public. From the table, it would appear that China has a reluctance to celebrate an individual scientist and that science in abstract is more popular, although it must be said that China has regularly published short sets of *Scientists of Ancient China* and *Modern scientists*. (An example of the two series are shown in Figures 4.12 and 4.13). The fact that most postal authorities do not show images of living persons places a bias on the number of scientists that might be chosen. Australia is an exception and since 1997 has issued an annual set of *Australian legends* stamps. From 1998, this has included live persons, culminating with the 2000, 2004, 2008 and 2012 Olympics, when it published a stamp depicting an athlete with a gold medal the day after the event. Subsequently, science achievements have had similar treatment. Five living medical scientists were represented on the *Australian legends* set of 2002. My study of the Australia Post files for this issue suggests that the chosen medical scientists were willing to be recognised in this medium and posed for the portraits used. A second series of living Australian *Medical scientists* was issued in 2012. The first living scientist to appear on the stamps of New Zealand was surgeon Sir Brian Barratt-Boyes, for a stamp in a set of *Famous New Zealanders* in 1995.

Living persons are generally not celebrated on postage stamps. The United States Post Office criteria include the statement that "No living person shall be honored by portrayal on US postage". Eastern Bloc countries have had a similar policy, although political leaders did appear, but in order to capitalise upon the success of the Russian space initiatives from the 1950s, they abandoned this policy and Yuri Gagarin became the face (not only on stamps) of that success on more issues than any other celebrant. In comparison, the US Postal Service celebrated its 1962 first space flight with a restrained image of a Mercury Space Vehicle, *Friendship 7*, but with no mention of pilot John Glenn, limiting the text to the words "US man in space". In 2000, included in the millennium celebration, there is a stamp showing the shuttlecraft *Discovery* with the words "return to space", but one had to read the narrative on the back of the stamp to learn that John Glenn had made a second flight at age 77. In 1994, the 25th anniversary of the first manned moon landing, a two stamp set showed two space-suited astronauts without naming them. So as not to miss out completely, during the cold-war space research propaganda race, the US issued several sets of stamps anticipating the future of space research, including the 1993 five stamp set titled *Space fantasy*. This set of images shows futuristic craft in space in order represent the science without reference to a specific scientist.

Figure 3.14 lists and represents the number of stamps that have been included within the study. The Russia, East Germany and Poland stamps are inflated by the number of space research stamps, a new category of research from the 1950s.

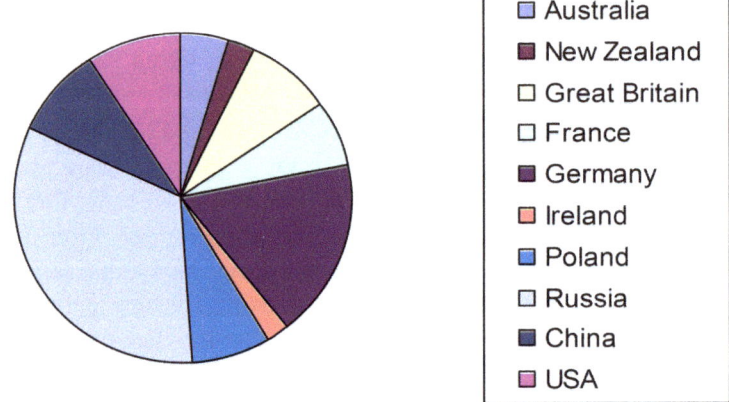

Figure 3.14: Pie-chart to show the ten countries considered within this study and their relative volumes against the total number of science stamps studied.

Source: Author's research.

Figure 3.15 charts the six main image classifications of my taxonomy.

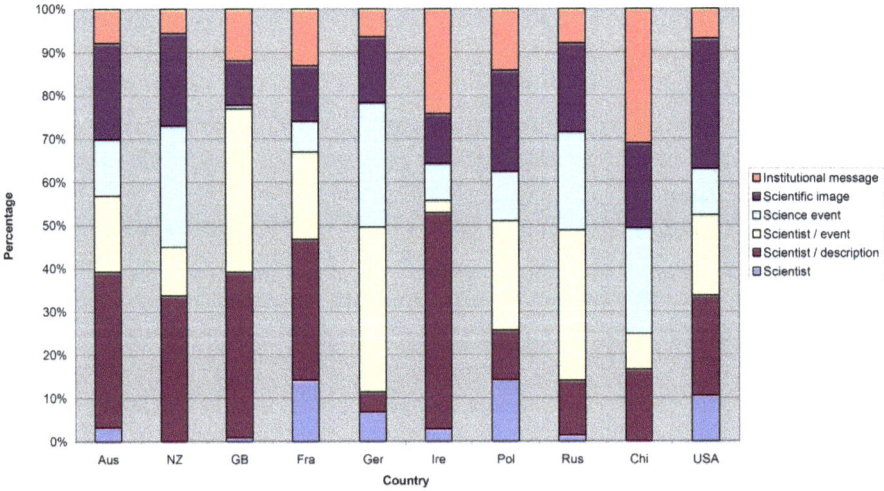

Figure 3.15: The breakdown of the science and scientist classifications over the total number of science stamps examined.

Source: Author's research.

Figure 3.15 enables the following observations to be made regarding the use of science and scientists on the stamps of the subject countries.

A named scientist with no further information comprise 5% of the total sample. I have combined two taxonomy descriptions into one category here. The results for a named scientist and a named scientist with life dates were both small.

France, Poland and the United States show a named scientist with no further elaboration, other than naming the subject, in 10% of their science issues. China and New Zealand have never issued a portrait of the scientist without any explanation of who he or she is or why they were chosen. Great Britain has issued just one stamp in this category, featuring Charles Darwin in one of a set of ten portraits from the *150th anniversary celebration of the 2006 (UK) National Portrait Gallery*. Australia, Ireland and Russia have issued very few stamps in this category.

A named scientist with a description of their achievement, in most cases features a symbol of the scientist's profession, or the subject is shown doing his or her job. These comprise nearly 18% of the total sample. Half of the Irish science stamps are in this category, however the number of Irish science stamps is low in comparison with the other countries, and is not significant in changing overall results. For Australia, New Zealand, Great Britain and France, one third of their science stamps are in this category, followed by the USA. Germany has issued very few stamps showing the scientist in the context of their work, just one in 20, seeming to prefer to recognise achievement on an anniversary date.

A named scientist celebrated in association with an event or an anniversary constitutes 27.5% of the total sample. Germany is the most prolific issuer of stamps in this category, constituting almost 40% of all the science stamps issued, followed by Russia and Great Britain. At the other end of the scale is Ireland. Again, this might be related to the small number of science stamps issued overall. New Zealand and China each have linked approximately 10% of their science stamps to an event or anniversary.

A science commemorated in association with an event or anniversary. Germany, New Zealand, China and Russia fall within the band of 22% to 29% of the total sample for these countries. Five of the remaining countries limit this category to about 10%. The notable exception is Great Britain, who feature a science image linked to an event or anniversary less than 1% of the time, preferring to include the image of a scientist through an association with an institution as parent of the event.

Scientific images comprise almost 20% of the total sample. The United States shows images rather than people for 30% of their science stamps. One reason might be that the United States Post Office decision not to show living persons on their stamps and to defer commemoration of anybody until a period after their deaths, which requires an alternative form of recognition to illustrate its message. Poland, Australia, New Zealand and China follow the United States in showing scientific images. Great Britain is the smallest contributor to this category, with 1 in 12 issued stamps being in this category. The reason for this is because Royal Mail concentrates on sending its messages through known scientists, often under an institutional umbrella, in order to describe scientific achievements or celebrate an event or anniversary

Institutional messages, celebrating an institution but also including public health campaigns or environmental issues constitute 12.4% of the total sample. China, followed by Great Britain and Ireland are the three authorities sending the most institutional messages. This is because these countries have tended to look at the big picture in celebrating international bodies. At the other end of the scale are New Zealand, Germany, United States, Australia and Russia, where the emphasis is on local issues.

Summary

The taxonomy developed for this study has enabled country comparisons and provided the basis for further analysis.

Another way of representing the taxonomy results is to show different countries' content practice by looking at the proportions of the stamps that describe a scientist or a science in abstract and compare those that celebrate an event or an anniversary to those with no time line. This analysis is shown in Figure 3.16. The main image and message on 50% of all the science stamps examined are personalised, leading to an axis labelled in the figure as "the human dimension". I have called the other 50% "science in the abstract". Another approximately 50/50 split, which forms the other axis of the figure, is between stamps with a defined date, generally a specific event or an anniversary, and stamps with no time frame implied (other than that of the known date of issue).

The four segments of the figure can be read to show that Russia's stamps, for example, are predominantly tied to a specific event or anniversary, with more named scientists celebrated than a science in abstract.

3. Classification and Analysis

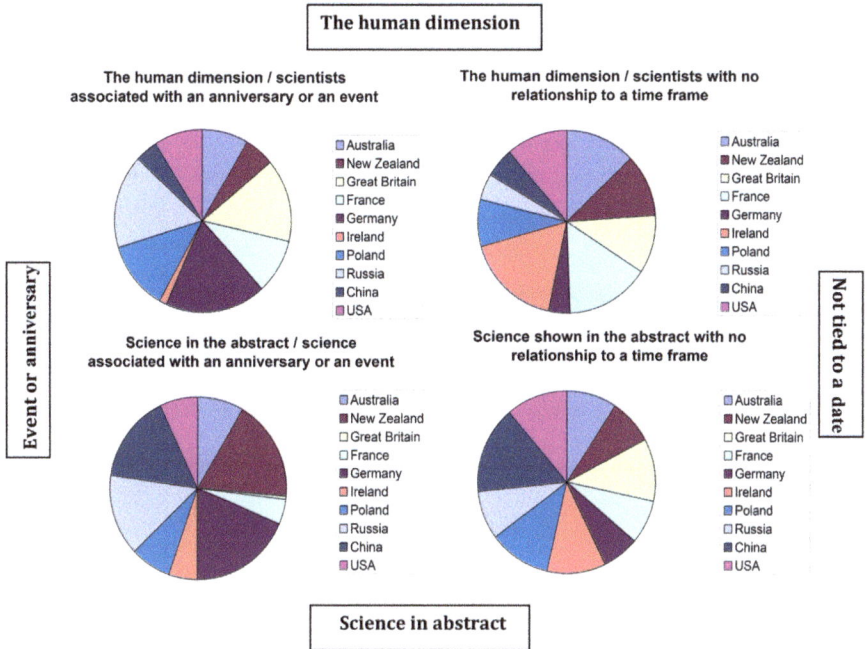

Figure 3.16: Representing the taxonomy in four quadrants.
Source: Author's research.

Interestingly each country shows a different taxonomy profile, reflecting its science history, amd the issuing authority's postal policy and political objectives.

The message on the stamp: A mirror or lens?

Is the message on a stamp a mirror of reality, or a lens, a view of new information or ideas? Posing this question offers the opportunity to ask how and why stamp images are chosen. Having examined the *Endangered species* issue and the *Action for species* response with the Head of Stamp Strategy of Royal Mail, it was appropriate to ask Mr Parker, Head of Stamp Strategy at Royal Mail in London, the same question. He said he considered awareness and perception to be elements of the design of all stamp issues. He also told me that, in authorising the concept and design of a new stamp issue, he needed to evaluate whether the image was reflecting something familiar to the general public, and thereby acting as a mirror of reality, or whether it was providing a view of new information or ideas, acting as a lens. We analysed the stamps issued by Royal Mail during 2010 and 2011 and were able to differentiate between mirror issues, such as the everyday definitive stamps showing the monarch's head, a design unchanged for 40 years, or commemorative stamps showing vinyl record covers and pictures from WWII, and a lens issue such as the 2010 celebration illustrate the achievements of the Royal Society at the time of its 350th anniversary. Mr Parker explained the process that had included consultation with the Royal Society in the choosing the ten scientists to be represented—one for each 35 years of the Society—and the interest it had invoked. I found it illuminating to be told that Alfred Russel Wallace had been chosen as the face of the theory of evolution, as Charles Darwin's work had been celebrated by Royal Mail with ten stamps the year before. So how do I engage with this set as a lens?

Figure 3.17: *The 350th anniversary of the Royal Society*. WNS catalogue # GB025–034.

Source: Author's collection.

As shown in Figure 3.17, the stamps are unusual. The bottom half of each each shows a black and white representation of the scientist, with the top half featuring a dramatic illustration of the designer's concept of what was in each scientist's mind, each of which features a different colour, "big minds with big ideas" (Davies, 2010, p. 15). The lower half of each stamp also features a one- or two-word description of the scientist's area of expertise, invention or area of study, with their family name shown at right angles to the basic design. The title of the stamp issue is shown across the descriptive image. Each stamp shows the monarch's profile and the service designator as '1st'. It is not obvious, without thinking about it, that each stamp represents a 35-year time period of the Royal Society. Collectively, the set tells the story of science from a Royal Society perspective as the science descriptors and the scientists reflect the era of their achievements: chemistry, optics, electricity, vaccination, computing, evolution, antiseptic surgery, atomic structure, crystallography, and earth structures. Each stamp summarises a complex subject visually with very little text. They are lenses.

In pursuing the concepts or mirrors and lenses, I use the term mirror to refer to a reflection of the familiar, a reinforcement of the status quo, a statement of what is known today, a representation of a national icon or history and a statement of today's reality. The lens, on the other hand, sends a message that informs or

3. Classification and Analysis

enlightens, and might require thought to decipher. A lens prompts the viewer to take a deeper look by showing something that might raise awareness of an issue and affect public opinion. The lens suggests behavioral change, prompting a level of interaction in response to the message it carries. After reflection, however, I have not incorporated lens or mirror into my formal taxonomy of stamps as shown and described in Figure 2.3. The distinction is too subjective and a stamp may, on occasion, have attributes of both mirror and lens. However, it is a useful prompt for analysing why particular stamps have been issued and that judgement is used for examples included in my study where appropriate. Prior to making this decision, I conducted two small surveys to test its validity.

The mirror or lens survey

I have discussed how I was introduced to the idea that the stamp image can act as a mirror or a lens and have used the concept to test the science message being told on a stamp throughout the results chapters. I have always believed the decision would be subjective. Surveys were conducted amongst colleagues and members of the general public. The surveys have been a worthwhile exercise confirming the subjectivity of the mirror or lens distinction.

I became very familiar with the process of classifying a stamp message as a mirror or lens. The concept was not so easily grasped by respondents.

Context is tied to history. Although I had dated the issue of the stamps upon which I was seeking respondents' judgement and given the stamp issue a title, it was not enough. Coming cold to the exercise was difficult. To give an example I showed respondents the Russian stamp of 1959 celebrating the *21st Congress of the Communist Party of the USSR* (shown in Figure 4.20). The image proclaims space research as a major achievement of the Party and is symbolically rich, showing the title of the congress, the country index, the value, and images of a moon rocket, Sputniks and satellites over a building that is flying the Russian flag, featuring the iconic hammer and sickle symbol. In the main, my responders were not aware of the significance of the date (the first Sputnik was launched in 1957) and historical context (particularly the importance of the space race between the USSR and the United States druing this time). I described this message as a lens but my respondents, less conscious of the importance of the date and historical context, declared it to be a mirror.

The three second opportunity to engage the viewer is a very small window of time. My colleagues who took the survey were interested to help and consciously spent time looking at each message in order to classify it as a mirror or lens. Our follow-up discussions indicated that they did not always seem able to decipher all the symbols, biasing their decisions.

I believe that the mirror or lens classification has helped me understand the messages on stamps for my study but the survey has confirmed to me that the subjective nature of any decision restricts its formal implementation.

Science on stamps: From the beginning

The first postage stamp (see Figure 3.18) was issued by Great Britain in 1840. It showed the image of the monarch, Queen Victoria, and self-described its role with the word "postage" and the value of the service being prepaid through its purchase. As the innovator of this fiscal device, there was no reason to state the country of origin and to this day Great Britain stamps use the portrait or an image of the current monarch as the index descriptor of the issuing postal authority.

Figure 3.18: Great Britain 1840. *The penny black,* (the first postage stamp). Gibbons catalogue # 1.
Source: Author's collection.

All other postal authorities are required by convention and, since 1874, the Universal Postal Union to state the country providing carriage of mail for the predetermined prepaid fee. Initially, succeeding images followed Great Britain's precedent, although the index might be a coat of arms of the country or a map, certainly something iconic at the time of issue.

One of the first authorities to issue a commemorative stamp portraying an image of the country was New South Wales, which celebrated the *Centenary of the First Fleet* in 1888.

Since then, all countries of the world have issued pictorial devices, particularly stamps that commemorate an event or an anniversary to show their own citizens, and anyone else who sees the stamp, the significance of the event.

The depiction of a science or a scientist took longer. It is relevant to consider the timing, the date of issue of the first scientists' images on postage stamps to help answer the following question: What does the representation of science and scientists on postage stamps convey about the political and cultural necessities of a country at the time of issue? Table 3.9 shows the date on which the first male scientist was honoured and highlights the fact that it took much longer for female scientists to be represented. Many countries, however, are still loath to portray a living person on their stamps. As previously stated, the United States Postal Service and the Citizens' Stamp Advisory Committee have as basic criteria for eligibility of subjects for commemoration: "No living person shall be honored by portrayal on US postage". This inevitably delays when a scientist can be featured on a stamp. As with every rule, there are exceptions, including Australia Post, who rush to celebrate sporting achievements and annually publish a set of living *Australian legends*. Great Britain tends to celebrate the institution rather than the scientist as has been described earlier in this chapter.

In the 170 years since the first postage stamp, the world has been through a scientific and technical revolution. The postage stamp has documented these new advances and acknowledged both the science and the scientists responsible. With regards to space research, of particular interest is the use of personality by Russia and Eastern Bloc countries to publicise their success in what became known as the space-race. The United States, on the other hand, with its self-imposed constraint of not showing living persons on postage stamps, issued many fewer and less dynamic stamps in comparison to communist countries. Alan Shepard, the first American in space, was not honoured on a United States stamp until the 50th anniversary of the event. Even the better remembered John Glenn has not been named on either of the stamps that celebrate his achievements. In contrast, Russia has issued more than 45 stamps showing images of Yuri Gagarin and naming him.

The first scientists shown on stamps

Table 3.9 lists the first male and female scientists to be celebrated. Illustrations of the stamps follow the table. I have noted both the first male and female scientists to be celebrated. In all cases, the female celebrant has been acknowledged several years after the first male. I discuss the representation of the female scientist on stamps in more detail later in this chapter. I shall look at controversy about claims by various countries to have been first to discover or invent achievements in Chapter Five.

Table 3.9: The first male and female scientists celebrated on stamps.

Country	Year	Male scientist celebrated / First female celebrated
Australia	1948	William J Farrer (1845–1906), agronomist and wheat plant breeder
	1970	Glenda Holloway,, CSIRO technician, *18th International Dairy Congress* (image only, Ms Holloway is not named)
New Zealand	1957	Sir Troby King (1858–1938), founder of Plunkett Society
		No female named scientist to date[1]
Great Britain	1965	Centenary of Joseph Lister (1827–1912), antiseptic surgery
	1970	Florence Nightingale (1820–1910), *1970 anniversaries*
France	1934	Death centenary of Joseph Marie Jacquard (1752–1834), inventor of Jacquard Loom
	1938	Marie Curie (1867–1934), 40th anniversary of discovery of radium
Germany	1934	Otto Lilienthal (1848–1896) and Count Ferdinand von Zeppelin (1838–1917)[2]
	1963	East Germany: Valentina Tereshkova (born 1937), cosmonaut
Ireland	1941	Sir William Rowan Hamilton (1805–1865), physicist
	2000	Marie Curie (1867–1934), *Millennium discoveries issue*
Poland	1923	Copernicus and Konarski, Polish anniversaries
	1947	Marie Curie (1867–1934), *Polish culture, celebrities*
Russia	1945	Mikhail V Lomonosov (1711– 1765), polymath
	1951	Sofia Koralevskava (1850-1891), mathematician[3]
China	1955	A set of four Scientists of Ancient China
	1980	Huang Daopo (13th century), *Scientists of Ancient China*
United States	1926	John Ericsson (1803–1889), screw propeller inventor[4]
	1940	Jane Addams (1865–1930), *American scientists*

Notes: 1) It appears that New Zealand maintains a chauvinistic approach to female scientists and their role in society. The bias has been introduced in looking only at named scientists. New Zealand has issued, on two occasions, generic female representations for *The International Year of Science* (1982), and shown a *Plunkett nurse* in the anniversary set of 2007.
2) Otto Lilienthal and Count Ferdinand von Zeppelin were featured in 1934 on specific air-mail stamps. Two years later the inventors of the motor car, Messrs Daimler and Benz, were celebrated. The third celebrant, Otto Von Guericke (1602–1686), established the physics of vacuums. His celebratory stamp (Figure 6.10) was issued during the period of the Third Reich and is discussed as a hero of science in Chapter Six. Earlier, during the Weimar Republic, the founder of the Universal Postal Union, Heinrich von Stephan (1831–1897), had been shown on stamps. Their contributions to the representation of science and scientists are included in my study.
3) In 1939, Russia issued a set of three stamps entitled *Women's Moscow–Far East flights* with the images of Polina Osipenko (1907–1939), pilot and hero of the Soviet Union, Valentina Grizodubova (1910–1993), pioneer aviatrix, and navigator Marina Mikhailovna Raskova (1912–1943). These are not shown in the table as the stamps are celebrating an aviation event rather than a scientific achievement.
4) In 1893, the United States issued a set of stamps celebrating Christopher Columbus. Within this book, feats of exploration by sea and on land have been, where appropriate, viewed as scientific achievements. But within the constraints of this table, the Ericsson memorial issue is more appropriate.
Source: Author's research.

The male and female Australian first scientists to appear on stamps are shown in Figure 3.19. Both stamps show a context for the issue. William Farrer's portrait is enhanced with a sketch of his wheat plants. The second stamp shows a generic woman laboratory technician at her CSIRO work-bench full of test equipment, it is fortuitous that we have learned her identity: Ms Glenda Holloway.

Figure 3.19: *William J Farrer, agronomist and wheat plant breeder* **and Glenda Holloway, CSIRO technician/***18th International Dairy Congress***. Renniks catalogue # 140 and 410.**

Source: Author's collection.

The first New Zealand scientists stamps are shown in Figure 3.20. The second stamp has a story of its own. It was printed to publicise the 1982 International Year of Science. "For reasons unknown this '*United Nations Year*' did not take place. The stamp is thus unique in 'celebrating' an event which never occurred" (Paterson, 2012, S62). The conference title is shown, as is the United Nations symbol. The scientist works at a control panel in a rural context with a microscope, circuit diagram and a communications satellite dish emphasising the scientific nature of the event.

Figure 3.20: Sir Troby King (1858–1938), Founder of *Plunkett Society* and A stamp celebrating the 1982 *International Year of Science*. Campbell Paterson catalogue # S74 and S282.

Source: Author's collection.

The Great British, French and German first scientist stamps are reproduced in Figures 3.21, 3.22 and 3.23. The final five countries first scientists, are shown in Figures 3.24–3.28.

Figure 3.21: Great Britain. *Centenary of Joseph Lister's discovery of antiseptic surgery*. Joseph, 1st Baron Lister (1827–1912) and *Florence Nightingale (1820–1910), 150th birth anniversary of this nursing pioneer*. Gibbons catalogue # 667–668 and 820.

Source: Author's collection.

Figure 3.22: France. *JM Jacquard (1752–1834), death centenary* and Pierre and Marie Curie, *40th anniversary of the discovery of radium*. Gibbons catalogue # 520 and 617.

Source: Author's collection.

Figure 3.23: Germany. *Otto Lilienthal* (1848–1896) and *Count Ferdinand von Zeppelin* (1838–1917), and Valentina Tereshkova (born 1937), cosmonaut, *Russian Space achievements Second Team, manned space flights*. Gibbons catalogue # 605 and E691–E692.

Source: Author's collection.

Figure 3.24: Éire. Sir William Rowan Hamilton (1805–1865), physicist and (although a foreigner) Marie Curie (1867–1934), *Millennium discoveries issue*. Hibernian catalogue # C22–C23 and C984.

Source: Author's collection.

Figure 3.25: Poland. *450th birth anniversary of Nikolaus Kopernicus* (1473–1543) astronomer, *150th death anniversary of Konarski* (educationalist) and Marie Curie-Sklodowska (1867–1934) scientist, *Polish culture*. Gibbons catalogue # 199–201 and 589–590.

Source: Author's collection.

Figure 3.26: Russia. *Mikhail V Lomonosov* (1711–1765), polymath, *220th anniversary of Academy of Sciences* and Sofia Vasilyevna Kovalevskaya (1850–1891), the first major female mathematician in the *Russian Scientists* set of 16. Gibbons catalogue # 456–457 and 1722.

Source: Author's collection.

Figure 3.27: China. Cheng-Heng (78–139) astronomer, Tsu Chung-Chi (429–500) mathematician, Chang-Sui (683–727) astronomer, Li Shih-Chen (1518–1593) pharmacologist and Huang Daopo, (13th century) textile expert from the *Scientists of Ancient China* sets of 1955 and 1980. Gibbons catalogue # 1660–1663 and 3024.

Source: Author's collection.

Jane Addams, shown in Figure 3.28, is the first female scientist to appear on an American stamp, appearing in the 1940 series featuring *American scientists*. Addams shared the Nobel Peace Prize of 1931 for her pioneering role as a social worker in America. It is interesting that a social worker has been chosen as a scientist, pre-dating the current initiative to include the humanities and social-sciences in science, as defined by *Inspiring Australia*, who refer to social sciences and the humanities as being "critical to the interface between science and society" (Department of Industry, 2010, p. ix).

Figure 3.28: United States. *John Ericsson* (1803–1889), screw propeller inventor, memorial and Jane Addams (1860–1935), social activist, Nobel Laureate from the 1940 *American Scientists* (set of four). Scott catalogue # 628 and 878.

Source: Author's collection.

Countries are keen to celebrate the achievements of their scientists. You will have noted that I have not included the early explorers who have been celebrated on the stamps of the United States (Christopher Columbus), or Australia and New Zealand (Captain James Cook). I have looked for the first local scientist, even though he may have been born elsewhere. The importance of the influence of Marie Curie-Sklodowska is evident in the fact that she is the first female scientist celebrated by three countries. Great Britain was the last country to celebrate individuals, other than members of the Royal Family, on its stamps. This policy change was made in 1965.

Women scientists on postage stamps

Having looked at the first female scientists to appear on stamps I note that, in addition to a recognisable scientist being shown on stamps, postal authorities have also used generic figures as images. Table 3.10 compares specific and generic images of women for the countries in this study. It would appear that generic female/worker images are more popular on the postage stamps of communist countries than capitalist authorities and in the United States, where living scientists cannot be featured.

Table 3.10: The number of stamps including images of named and generic women scientists.

Country	Named female celebrant	Generic female / worker image	Total
Australia	6	2	8
New Zealand	-	3	3
Great Britain	4	3	7
France	3	3	6
Early Germany	-	-	-
East Germany	3	14	17
West Germany	4	-	4
Re-unified Germany	-	1	1
Ireland	3	-	3
Poland	8	-	8
Russia	5	4	9
China	2	7	9
United States	17	2	19

Source: Author's research.

Generic figures are used by postal authorities when they need an aspirational model. Classically, it will be a worker shown in the context of a working environment, with the model contributing to the general good of the country. Figure 3.29 is a good example of a stamp with a political message, and is of special interest to the collector of stamps as the stamps are not perforated for easy use, the expectation is that the whole miniature sheet will be used.

Figure 3.29: East Germany, 1963. A generic female (laboratory) worker within the miniature sheet titled *Chemistry for Freedom and Socialism*. Gibbons catalogue # MSE674a.

Source: Author's collection.

The first female medical practitioners

Four of the ten countries being considered within my study have issued stamps to honour their first female medical practitioner. Australia has celebrated Constance Stone twice. The second stamp in Figure 3.30 includes the portrait and two scenes recording the female doctor at work. Figures 3.31, 3.32 and 3.33 show images with limited text.

Figure 3.30: Australia, 1975 and 1990. One stamp from a set of six famous Australian women, *Constance Stone the first woman medical practitioner* and the celebration of the *Centenary of Women in Medical Practice*. Renniks catalogue # 550 and 1172.

Source: Constance Stone.

3. Classification and Analysis

Figure 3.31: Great Britain, 2008. Dr Elizabeth Garrett Anderson, first woman to earn a medical qualification, one of a set of six *Women of Distinction*. Gibbons catalogue # 2871.
Source: Author's collection.

Figure 3.32: West Germany, 1987. Dorothea Christiane Erxleben, *First female medical doctor*. Gibbons catalogue # W2155.
Source: Author's collection.

Figure 3.33: United States, 1970. *Great Americans* definitive: *Dr Elizabeth Blackwell, first woman physician in the US*. Scott catalogue # 1393D.
Source: Author's collection.

Introduction to the case studies

In this chapter, I have examined the results of my taxonomy and made an initial analysis of results. To further examine the theme, a number of case studies have been completed. Each country has been reviewed from oldest to newest issues relevant to the topic of the case study. In simple terms this has illuminated the opportunities embraced by stamp designers to enhance the message the stamp is carrying. When looking at a number of stamps, logically of the same subject, it is possible to view how the message that is presented has changed. A multiple case study approach has been taken and each question analysed against the findings of each case study to be able to draw relevant conclusions.

Different perspectives were chosen to broaden the base of my study in the expectation that additional narratives would emerge.

I am interested to explore if politics influences the message on the stamp. After establishing the basic results for the study, this became the first case study (Chapter Four). Following on from this premise, a look at the specific stamps of the Antarctic Territories of four of the prime countries seemed obvious. This led into looking at how other countries reflected their interest in Antarctica on their own stamps.

The second area for case studies has been to look at scientists on stamps (Chapter Five).

The celebration of firsts claimed within the message contained in a stamp, in simple terms, and to test Merton's thesis is that it is the institution that claims priority of discovery rather than the scientist..

The third area for case studies is looking at which scientists appear as heroes of science, through an examination of the number of times their achievements have been used to tell a message (Chapter Six). Do countries celebrate the achievements of foreign scientists and does this reflect, in any way, an awareness of science?

Stamps issued at the turn of the millennium to celebrate that event have also been studied as an entity (Chapter Seven). The countries that have issued millennium stamps have evaluated the impact of science of the twentieth century upon their ways of life. Out of this study, the surprise that environmental issues were seemingly absent in the minds of the postal authorities caused me to look, as another time related opportunity, at how the changing climate has been reflected on postage stamps.

4. Stamps as Communicators of Public Policy

> Every state is engaged ... in using [stamps] as a propaganda vehicle (Stoetzer, 1953, p. 3).

I have identified postage stamps conveying a science message from ten countries. This allows me to examine the development of these images over time to explore the following questions: What messages did governments and postal authorities wish to convey at what times? Have specific events influenced the messages being sent? These questions are pointers to the first research question in this study: What does the representation of science and scientists on postage stamps convey about the political and cultural necessities of a country at the time of issue? As Stoetzer notes: "Today governments are unabashedly using postage stamps to promote domestic products, vacation resorts, cultural achievements, and even political ideologies" (Stoetzer, 1953, p. 3).

The newest countries in my study illustrate public policy on stamps the most comprehensively. These one-party states have adopted a stamp policy related to building national identity, and formulating and advertising plans to use science and technology to develop strategies to improve the lot of the general public. The proposition is clearly demonstrated by China, Russia and East Germany from 1949 to 1990. The fact that these are, or were, communist regimes is not mere coincidence, because the totalitarian state uses postage stamps as potent messengers of policy. After WWII, Germany was divided into two countries, each pursing a different ideology. A comparison of the postage stamps issued by these countries and their use of science and technology to promote those ideologies is reported in this chapter. Observations will be made to test the mirror or lens idea, in addition to comments on the semiotic approach used within the design of the images.

Established democratic states do not send overt political messages but as we shall see there are national issues that are raised, on stamps, to prompt behavioural change through policy related to health care and, currently, the changing climate. Science and technology messages serve as statement of benefits historically and in relation to everyday living.

Also included within this chapter is an examination of the stamps of Antarctica, through which the countries of the world send messages concerning their continuing interest in this region.

As Collins and Pinch observe in *The Golem: What you should know about science*, science and politics are closely interrelated, with scientific and technological issues increasingly figuring in the political process and the mass media affording a public understanding of the political role of science and technology (Collins and Pinch, 1998, pp. 142–143). An analysis of the messages on stamps, as mass media, enables such an understanding.

The People's Republic of China

China in particular lends itself to an analysis of stamps as communicators of public policy. China has existed in its present form as a communist state since 1949, and its overt policies have been clearly enunciated by the leader of the day. The world has watched the successes and failures of the policies adopted by China, and one source of information has been the messages conveyed on postage stamps, permeating the day to day experience of the users of the postal service. After the communist victory in 1949, efforts were made to organise science and technology in China based on the model of the Soviet Union. Figure 4.1 shows a teacher, an older Chinese man, demonstrating the three phases of agriculture to two students: manual labour with human driven tools, the use of animal power, and, dominating the image, a mechanical tractor. The stamps implies that mechanism is coming to the land and that life will be easier as a result. The text is minimal and each of the four stamps is printed in a single colour. The value of the stamps varies from 100 fen to 800 fen, indicating that this is a message intended for all classes of mail and all possible destinations. It is a mirror of reality to the politician, but possibly a lens to the future for agrarian communities. For those outside China, the symbolism will be understood by the older figure pointing to the future, a direction formulated by the modern tractor.

Figure 4.1: China, 1952. *Agrarian reform*, from a set of 4 stamps with the same design. Gibbons catalogue # 1352.

Source: Author's collection.

During the first ten years of the People's Republic of China, homage was paid to Russia, with images of Russian leaders such as Stalin, Marx and Lenin appearing on Chinese stamps. One such image shows Mao Zedong in conversation with Stalin. In 1955, a few months prior to the inauguration of a new Five Year Plan, a set of four stamps was issued celebrating *Scientists of Ancient China*. These were shown in Chapter Three, Figure 3.27 represents the first scientists recognised by the People's Republic of China. The set celebrates the four scientists whose portraits dominate the image, their achievements are shown in text but it is the scientist who is on show. The message is also that the People's Republic of China will build upon the achievements of the past.

During the period 1958–1961 the Communist Party of China inaugurated its policy of reform known as the *Great Leap Forward*, and used stamps to promote it, as is shown in Figure 4.2. A national exhibition was celebrated with the issue of these three stamps. The symbolism is Chinese and clear: from left to right, emblem and exhortation, dragon over clouds analogy for "aiming high", and flying horses to represent the Great Leap Forward. The stamp value reflects the currency revaluation of 1955 and the common value for this set indicates the message is for in country dissemination through local mail service. The stamps show representations of science and technology at the base of the image. I believe these are lenses, looking towards Mao's better future.

Figure 4.2: China, 1958. *The National Exhibition of Industry and Communications.* **Gibbons catalogue # 1782–1784.**
Source: Author's collection.

China showed its closeness to the Soviet Union in 1958 by issuing a set of stamps to celebrate the launch of the first Russian Sputnik, and in 1959 a set celebrated the launch of the first lunar rocket.

A second set of four *Scientists of Ancient China* was issued in 1962, a set in advance of its time, as each scientist is shown both as a portrait and in context of their disciplines: an inventor (Tsai Lun ?–121), a physician (Sun Szu-Miao (581–682)), a geologist (Shen Ko (1031–1095)) and an astronomer (Kuo Shou-Chin (1231–1316)).

Figure 4.3: China, 1962. *Scientists of Ancient China*. Gibbons catalogue # 2055–2062.

Source: Author's collection.

Mao Zedong's "Theory of Productive Forces" was a disaster and the *Great Leap Forward* resulted in millions of deaths through starvation during the great Chinese famine. The Communist Party of China criticised Mao, leading him to initiate the Cultural Revolution in 1966. Being an intellectual was a dangerous occupation, as Mao and Jiang Qing organised students known as the Red Guards to spread their ideology across the country. Mao Zedong's Cultural Revolution of 1966–1969, and its aftermath, which lasted until 1976, had a catastrophic effect on Chinese research, as academics were persecuted and the training of scientists and engineers was severely curtailed for nearly a decade (Tsou, 1968).

China issued no science stamps during the 11 years of the Cultural Revolution. People's Republic of China stamps generally used images of the population enthusiastically following the advice of Mao's *Little Red Book*. Many stamps were issued showing nothing more than text praising Mao, or his own words. An example is shown in Figure 4.4. It is a significant stamp within this study and proves the importance that the issuing authority places upon the accuracy of its images in stating its message. This stamp, although titled *The whole country is red*, was issued and immediately withdrawn from sale when authorities noticed that the island of Taiwan, which the People's Republic of China has always claimed to be a part of China, was shown in white.

4. Stamps as Communicators of Public Policy

Figure 4.4: China. *The whole country is red*. Gibbons catalogue # 2403a.
Source: Author's collection.

The number of science stamps issued by China is recorded in Figure 4.5. This figure highlights how the implementation of the Cultural Revolution changed the pattern of the number of science stamps being issued by People's Republic of China, marked as (a). There was also a period of three years 1993–1995, marked (b), when science images were not used on Chinese stamps. It is significant that this gap followed the Eighth National People's Congress in 1993, and the election of Deng Xiaoping, whose policies were to move to a more market-driven economy, with science perhaps put on a temporary back-burner. I note that I could not find a similar condition after the 2002 election of Jiang Zemin. This could be a result of the trend of the People's Republic of China to issue more science stamps after the year 2000, noted on the chart. Also highlighted are two years, 1978 and 2002, when the People's Republic of China issued sets of stamps to convey science message rather than single stamps. These years may be compared with 1982, when science was presented in single stamps. The red arrow to the right of the chart marks an increase in the number of science stamps being issued after 2000, which was noted in Table 4.5.

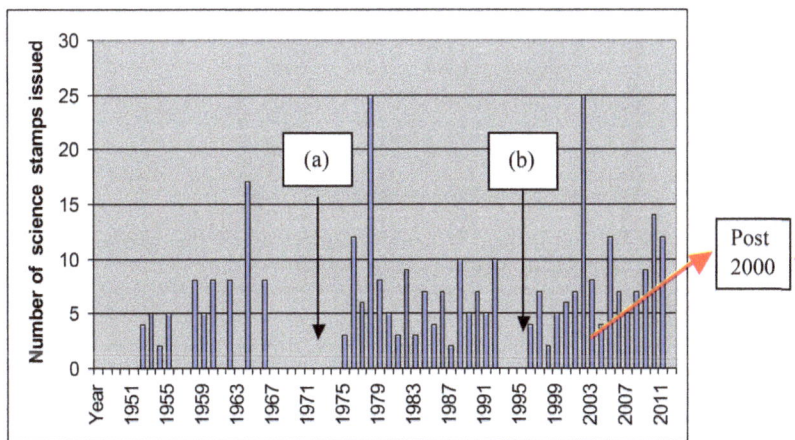

Figure 4.5: The dearth of science stamps issued by China during the Cultural Revolution. Also highlighted are two years when the People's Republic of China issued sets of stamps to convey messages rather than single stamps.
Source: Author's research.

After Mao's death in 1976, science and technology were established as one of the "Four Modernisations" and China's Soviet-inspired academic system was gradually reformed. The period 1976–1989 is described as the "Era of Reconstruction", during which science was used to promote images of reconstruction (Tang Tsou, 1968).

In 1976 there were three sets of stamps with a science theme, mirrors illustrating what was being achieved by the State through science and technology.

1. *Medical Services*, a set of four. Interestingly, all four images include an operation with acupuncture anaesthesia, one image of an actual operation and three that show patients in recovery with the original operation appearing in the background.

2. A celebration of the *Success of the proletarian education system*, a set of five. Four of the images are of happy students and workers, but one features a computer, for the first time, on a Chinese stamp (Figure 4.6). This stamp is possibly a lens, requiring explanation of a console, printer, paper tape and magnetic tape units. The two main women operators are shown wearing white coats, suggesting the importance placed upon the role of women as part of the modernisation of science and technology. The white coat can also be read as a semiotic symbol for science.

3. *Maintenance of electric power lines,* showing four aspects of electrical service and technology.

Figure 4.6: China, 1976. An image from the *Success of the proletarian education system* set. Gibbons catalogue # 2666.
Source: Author's collection.

In direct contrast to these science-themed images, 1976 saw the issue of a set of three featuring the tenth anniversary of *Mao's May 7 directive*, In this directive, Mao had suggested setting up farms, later called cadre schools, where cadres and intellectuals, sent from the cities, would perform manual labor and undergo ideological reeducation (Tang Tsou, 1968).

The following year, 1977, a set showing *Developments of the Petroleum Industry* was released, with six designs illustrating the development of technical infrastructure for the industry, such as drilling rigs, pipe-lines, port facilities and off-shore drilling.

1978 proved to be a bumper year for the promotion of the importance placed upon science and technology with seven issues, in sets, featuring the development of different industries. The issues all tell a story, probably introducing new technologies. They therefore might be considered as a specific lens on the future. The stamp denominations are all local service.

1. *National science conference*, a set of three stamps rich with scientific icons, including the atomic symbol, the globe and space orbits.

2. *Meteorological services*, a set of five (see Figure 4.7). The images are described as illustrating, from left to right: *Releasing weather balloon, Radar station: typhoon watch, Computer weather maps, Local weather observers* and *Rockets intercepting hail clouds*. These stamps are highly contextual and show the interface between the technology and the operatives of the technology.

3. *Chemical industry development*, a set of five containing symbolic representation of fabric production.

4. *National conference on agriculture: learning agriculture from Taichai*, a set of three illustrations of on-the-job training focusing on progress in animal husbandry.

5. *Iron and steel industry*, a set of five illustrating the heavy engineering infrastructures in the manufacture of steel.

6. *The Ninth National Trade Union Congress*, a single stamp. I would not generally classify this as a science subject if it weren't for the fact it contains the atomic symbol, a cogwheel, grain production and a rocket.

7. *Mining development*, a set of four designs with images of mineralogical discovery and mining infrastructure.

Figure 4.7: China, 1978. *Meteorological services*. Gibbons catalogue # 2766–2770.

Source: Author's collection.

Then followed a somewhat barren period for science, although stamp issues celebrated the achievements of two foreign scientists: in 1979 the birth centenary of Albert Einstein, and in 1982 the centenary of the discovery of the tuberculosis bacillus by Dr Robert Koch, scientists acclaimed for their achievement by many countries of the world. With these issues, China acknowledges its place in the world

In 1980, China issued a single stamp that has strongly influenced my study, and which was introduced in Chapter One. The stamp announces the Second National Conference of the Chinese Science and Technology Association (see Figure 4.8). Just four years after the effects of the Cultural Revolution, during which time science had no place, China promoted a science and technology conference with a stamp that is semiotically rich (as discussed in Chapter Three). It is a long-focal-length lens that use the past as a path to the future. The postal authority used asparas again in 1983 to tell a similar message for World Telecommunications Year. This is shown within Figure 4.8 for comparison.

4. Stamps as Communicators of Public Policy

Figure 4.8: China, 1980 and 1983. *Second National Conference of the Chinese Science and Technology Association* **and** *World Communications Year*. **Gibbons catalogue # 2974 and 3247.**
Source: Author's collection.

From the 1980s, there was a change in the way that science was promoted and Chinese stamps characterised an authority that was looking outwards to the world with single stamp issues. This next series of messages are lenses, as they challenge the Chinese general public to think outside of local issues.

1. 1981, *World Telecommunications Day*, this single stamp is visually rich, incorporating the symbols of the International Telecommunications Union and the World Health Organisation, joined by a multi-coloured caduceus, a recognised symbol of commerce and negotiation (see Figure 4.9).

2. 1982, *World Food Day*, a single stamp showing the logo of The Food and Agriculture Organization of the United Nations, the globe, racial profiles, and an ear of wheat.

3. 1982, *The United Nations International Drink Water and Sanitation Decade* dedicated to helping people escape the poverty and disease caused by living without safe water and sanitation symbolised by dripping water and children reaching towards the source of the drip.

4. 1982, *Second United Nations Conference of Exploration and Peaceful Uses of Space*: *Unispace 1982,* identified with the conference symbol, the earth as a globe connected to another planet in space.

5. 1983, *World Communications Year*, represented by winged messengers (asparas) and a globe, as described above and shown in Figure 4.8.

Figure 4.9: China, 1981. *World Telecommunications Day*. Gibbons catalogue # 3071.
Source: Author's collection.

From 1986, China publicised its entry into space research with classical images of satellites, rockets and orbital representations (see Figure 4.10). The stamps tend to be larger than other issues and have limited text. Previous space issues had been celebrations of Russian successes such as the first Sputnik in 1958, and the lunar rockets of 1959 and 1960. US successes were not recorded on the stamps of China.

Portrayed as a triumph for Chinese science and technology and a milestone for Chinese nationalism, the launch of the first Chinese manned space craft was celebrated in 2003. China became the third country to put a man into space: Yang Liwei. The space-suited (without helmet) cosmonaut is celebrated on one stamp saluting the red Chinese flag and fully-suited cosmonaut on another has a space-craft, in space, above his left shoulder. They are mirrors of the actual achievement.

The 50th anniversary of the Chinese space programme showed the images of geospace double star exploration and the Shenzou-VI manned space ship (see Figure 4.11). The style of the images indicates a telescopic lens looking into the future.

4. Stamps as Communicators of Public Policy

Figure 4.10: China, 1986. *Space research*. Gibbons catalogue # 3423–3428.
Source: Author's collection.

Figure 4.11: China, 2006. *50th anniversary of the Chinese space programme*. Gibbons catalogue # 5094–5095.
Source: Author's collection.

One series that was continued was the celebration of the *Scientists of Ancient China*, issued in 1955, 1962, 1980 and 2002. There is no attempt to illustrate context other than through text, unlike the series shown in Figure 4.3, which showed the scientist going about their trade. Three of the stamps are for local service, but the 60 fen denomination, featuring a female scientist, is for international service. The scientists are agronomist Xu Guangpi (1562–1633), hydraulic engineer Li Bing (3rd century BC), agronomist Jia Sixie (fifth century), and textile expert Huang Daopo, (13th century). From 1988, the emphasis was changed to a series of *Modern scientists,* an example is shown in Figure 4.13.

Figure 4.12: China, 1980. *Scientists of Ancient China*. Gibbons catalogue # 3021–3024.

Source: Author's collection.

The images for the *Modern scientists* series continue to feature the scientists, but include a meaningful contextual illustration to acclaim their expertise, providing a lens into what would otherwise be mirrors celebrating the scientist. Those featured are gynaecologist/obstetrician Lin Qiaozhi (1901–1983), astronomer Zhang Yuzhe (1902–1986), chemist Hu Debang (1890–1974), and agronomist Ding Ying (1888–1964). The values are for local and regional service.

The set of five stamps issued by China for the millennium are discussed in Chapter Seven. Only one of the five images incorporates a scientific message (Figure 7.12). Environmental protection provided the subject matter for four issues in 1988, 2002, 2004 and 2010, and is discussed in some depth in Chapter Seven.

In today's China, science and technology are seen as vital for achieving economic and political goals, and are held as a source of national pride to a degree sometimes described as "techno-nationalism" (Kang and Segal, 2006). Relevant stamp issues cover major projects such as the 2001 *Construction of the Qinghai-Tibet railway,* the controversial 2002 *Hydroelectric power generation and water control on the Yellow River*, and the 2005 *Tarim-Baihe gas pipeline*. The Yellow River achievement commemoration is reproduced in Figure 4.14. This set is the first examined from China where photography provides the image. These are mirrors, the tangible reflections of huge construction projects.

4. Stamps as Communicators of Public Policy

Figure 4.13: China, 1990. *Modern scientists*. Gibbons catalogue # 3702–3705.
Source: Author's collection.

Figure 4.14: China, 2002. *Hydroelectric power generation and water control on the Yellow River*. Gibbons catalogue # 4722–4725.
Source: Author's collection.

Summary of the science stamps of China

The People's Republic of China is a relatively new country. Its political leaders, especially Mao Zedong, have had a profound effect upon the country's aims and ambitions and have publicised their expectations on postage stamps. Science messages comprise 8.4% of the total issue from the country. Political policy was not articulated, after Mao, by edicts on stamps but through images of workers enjoying the advantages brought by science and technology. Context has been important with these images to make sense of the message. Science themes have been developed within sets as well as single stamps, except when China looks outward to the world to celebrate a worldly event such as the issues of the early 1980s, as discussed above.

Scientific heritage has not been forgotten with particular attention paid to the *Scientists of Ancient China*, with images celebrating the individual rather than the achievement, the message being that science is not new to the country's development. The new series, *Modern scientists*, illustrates the person with a sympathetic portrait and an image of their area of study. Later images have incorporated photographs to tell the message.

This analysis clearly supports Stoetzen's proposition of stamps as a propaganda vehicle. In particular, the significant use of science on stamps supports the notion of techno-nationalism.

Soviet Russia and East Germany

Two other countries provide significant examples of the use of science to promote national plans or projects. These are Soviet Russia (USSR, 1923–1991), which contracted to the Russian Federation in 1992, and East Germany, which existed as a constitutional entity within the Eastern Bloc in Europe from 1949–1990. The Russian postal statements are of a larger format than normal and are somewhat monumental in the use of the images drawn for the issue. Photographs are rarely used.

Russia

The stamps of the two earliest stamp issuing Russian constitutions, the Empire of all the Russias (1855–1917) and the Russian Socialist Federal Soviet Republic (RSFSR, 1917–1923) featured a mixture of patriotic images with generic representations of agriculture and industry, most usually featuring a hammer and sickle. The USSR was formed in 1923. The first stamp issued by the USSR celebrated the 1923 Agricultural Exhibition in Moscow and showed images of generic reapers and sowers using hand tools. The first science and technology image was also included. It shows a tractor (see Figure 4.15), acknowledging that Russia purchased 25,000 Fordson tractors from the Ford Company in the US between 1921 and 1927. In 1924, the Leningrad plant "Red Putilovet" (Красный

Путиловец) started production of Fordson-Putilovets tractors (Фордзон-путиловец), for home consumption (Gatrell, 1994). The stamp shows the tractor almost enveloped by the figure 5 (indicating a Five Year Plan) and growing hop vines. A scene of hills is visible under the tractor. The stamp is a lens, bringing into focus the introduction of science and technology to Russia.

Figure 4.15: Russia, 1923, *Agricultural exhibition*. Gibbons catalogue # 327.

Source: Author's collection.

After 1953, in the post-Stalin years Communist Party platforms continued to occupy a prominent place on Soviet stamps, but were presented in a different manner. Gone were the brief heroic slogans of the Stalinist era that urged economic mobilisation, and in their place were rather lengthy excerpts from Party Congresses. For example, in 1962, one stamp in the series of *Decisions of the XXII Congress of the Communist Party of the Soviet Union – into life* proclaimed that "by 1980 livestock, cattle and chickens will be significantly increased. Production of meat will grow almost four times, milk almost three times" (see Figure 4.21).

The science stamps of the USSR until 1959 established a trend for the country to celebrate an individual scientist's achievements mainly on birth or death anniversaries. Lomonosov and the Leningrad Academy of Sciences, and Popov, as the inventor of radio are the predominant individuals honoured. Exploration and the developing aviation industry feature in sets. An *Aviation day* is celebrated with long sets of 9 stamps of various designs.

I regard a 1951 set of 16 stamps with the title of *Russian scientists* as significant, almost as if the issuing authority decided it was time to celebrate individual scientists. Within the set is the first female scientist to be honoured, Sofia Vasilyevna Kovalevskaya (1850–1891), recognised as the first major Russian female mathematician. The other five images shown in Figure 4.16 include a contextual indication of the scientists' professions and are among the earliest stamps to do this. The remaining stamps are typical of the poster style of Russian stamps, featuring an individual and minimal text without context, a style still used today.

The Representation of Science and Scientists on Postage Stamps

From left to right, the six scientists shown are:

1. Pavel Nikolayevich Yablochkov (1847–1894), an electrical engineer.

2. Nikolai Alekseevich Severtzov (1827–1885), explorer and naturalist.

3. Konstantin Eduardovich Tsiolkovsky (1857–1935), rocket scientist and pioneer of astronautic theory.

4. Alexander Nikolayevich Lodygin (1847–1923), electrical engineer and inventor, according to Russia, of the incandescent light bulb.

5. Dmitri I Mendeleev (1834–1907), chemist and inventor.

6. Sofia Vasilyevna Kovalevskaya (1850–1891), mathematician.

Tsiolkovsky, Mendeleev and Kovalevskaya are also featured on other Russian stamps.

Figure 4.16: Russia, 1951. *Russian scientists*. **Gibbons catalogue # 1710, 1714, 1715, 1716, 1720 and 1722.**
Source: Author's collection.

Although these stamps feature contextual indications as an aid to understanding the importance of the scientists, the context is greatly overshadowed by the faces. The person is the focus here, not the work, and it is clear that the stamps are intended to emphasise Russia's scientists as individuals achieving greatness on an international stage. The stamps are intended to inform the public of the achievements of Russians, and constitute more of a lens than a mirror of public awareness.

The complete set of 16 are all the same value, a low service value suggesting the message is intended for use internal to Russia, partly refuting a Merton's observation: "The recent propensity of the Russians to claim priority in all manner of inventions and scientific discoveries thus energetically reduplicates the earlier, and now less forceful though far from vanished, propensity of other nations to claim like priorities" (Merton, 1957, p. 642).

The USSR has continued to issue stamps commemorating individual scientists. These are discussed further in Chapter Six. In this chapter, however, I shall focus on the importance and the implications of stamps featuring science rather than scientists.

In 1959, the USSR issued its first overtly political message with a set of 12 stamps with a large figure 7 representing the Communist Party Seven Year Plan. The images are of a variety of manual workers and infrastructure, but one stamp shows a generic, female chemist (see Figure 4.17). The chemist is working under a forceful figure 7 containing an upward arrow proclaiming a 300% target rise in chemicals production between 1958 (shown as the 100% start point) and 1965. The white-coated chemist is at her desk, holding laboratory equipment and looking out on a major chemical work as she reports. This is most certainly a lens, with a very strong political application of the science of chemistry.

Figure 4.17: Russia, 1959. *Chemist* from the Seven Year Plan. Gibbons catalogue # 2360.
Source: Author's collection.

In 1957, Russia announced to the world that it had launched the first artificial satellite, celebrating the achievement with a stamp issue. A second artificial satellite launch was commemorated the same year with a set of four stamps, *To the stars*. These celebrations are reproduced in Figure 4.18. A symbolic female figure, from a sculpture of Ye Vuchetich, is shown standing upon a globe, the earth, gesturing towards the vapour trail of the rocket making its way to the stars. A building topped by a star-shape denotes the state's endorsement of the venture. Four different values were issued, possibly as a short definitive set, implying Russia expected a longer sales life for this commemorative issue.

In this case, the news of Russian scientific achievement is given to the public through a familiar and romantic icon, which is exploited in a scientific context to convey a strong political message. The stamp mirrors Russian public pride in the event, while at the same time seeking to raise nationalistic feelings. With a range of service values, it would be expected that the message would travel outside of Russia.

Figure 4.18: Russia, 1957. *Launches of the first and second artificial satellites*. Gibbons catalogue # 2147–2418 and 2164–2167.

Source: Author's collection.

In 1958, the third artificial satellite was successfully launched. Figure 4.19 shows the stamp and an explanatory note, se-tenant (adjacent with a different design), throughout the printed sheet of stamps (Scott Carlton, 1997). This stamp represents a marked change from the previous more subtle styles of Figure 4.18. The public is "being educated" here in an uncompromising manner. There is no room for misunderstanding this message about Russian science; it is being trumpeted for all to see and read through a nationalistic lens.

4. Stamps as Communicators of Public Policy

Figure 4.19: Russia, 1958. *Launch of the third artificial satellite*. Gibbons catalogue # 2222.

Source: Author's collection.

In 1959, the first launch of a moon rocket was celebrated under the banner of *21st Congress of the Communist Party of the USSR*, one of a set of three (Figure 4.20). This stamp proclaims space research as a major achievement of the Party, it is symbolically expansive, showing the title of the conference, the country index, the value, and the images of a moon rocket and Sputnik satellites over a building flying the Russian flag. The iconic hammer and sickle symbol is also shown.

Figure 4.20: Russia, 1959. *21st Congress of the Communist Party of the USSR*. Gibbons catalogue # 2302.

Source: Author's collection.

It is of interest that Figures 4.18–4.20 present space achievements as a political end in itself. This is in sharp contrast to Figure 4.17, where the science of chemistry is being used to promote a broad political goal with an industrial emphasis rather than a scientific one.

The first and second manned space flights, of Yuri Gagarin and Gherman Titov respectively, took place in 1961. The stamps featuring images of the cosmonauts and their vehicles are discussed in Chapter Five, as part of the examination of people whose achievements underlay the messages being relayed. Space research stamps from Russia and Eastern Bloc countries appeared regularly from this date and continued for almost 30 years.

Russia also features its science and technology aspirations and achievements through the publication of Communist Party Conferences and specific industry reviews. One such example is shown as Figure 4.21. All nine of the images in this set show production targets to be achieved by 1980. From left to right, the images in Figure 4.21 represent: the chemical industry and statistics, hydro-electric power, telecommunications, and heavy industry. It is worth noting that two of the four science images include a female worker. Once again, the signs are of science and technology, but they signify industrial progress. Science and technology are, in a sense, being used here to reinforce the political will and drive national aspirations. The extensive text in Russian suggests that the stamps were intended for local rather than international service.

Figure 4.21: Russia, 1962. *Achievements of the people, 22nd Communist Party Congress – Into life*. Four from the set of nine (large stamps) that feature a science. Gibbons catalogue # 2766–2774.

Source: Author's collection.

This is also the message within the eight stamps of the 1965 *Industrial progress* issue (Figure 4.22). Unlike some of the other Russian stamps reviewed, these are definitive stamps covering a range of pre-payments for postal service. Previous political messages, it would appear, are intended for local use and their denomination pre-pays in country mail fees. Figure 4.22 however, shows the images of a power station, steel works, chemical works and formula, machine tools production, building construction, agriculture, communications, and transport with the highest value item showing scientific and space research.

The purpose of these stamps, issued with a range of service prices and minimal Russian text, therefore, is to indicate to the world outside Russia that industrial progress is taking place across a number of scientific disciplines. Future goals are nowhere to be seen, only the present. The stamps are clearly intended as a lens into Russia at that time, demanding respect for her science and technology.

Figure 4.22: Russia, 1965. *Industrial progress*. Gibbons catalogue # 3166–3173.

Source: Author's collection.

The 1966 definitive set of 12 Russian stamps (Figure 4.23) includes many references to science and technology as a part of Russian life. Half of the value issues are promoting science. The hammer and sickle symbol appears regularly. Generic workers emphasise how scientific and technological development has influenced their lives. Each stamp contains a symbol of the state. In smaller letters the the date of issue, 1966, is shown, indicating the 50th anniversary of the state.

Figure 4.23: Russia, 1966. Definitive set. Gibbons catalogue # 3347–3358.

Source: Author's collection.

The 1966 set was reissued in 1968 with the same images, although there were some colour changes, so the same set of images would have been in circulation for a number of years.

The year 1967 saw the 50th anniversary of the October Revolution that led to the establishment of the USSR. Ten large stamps were issued, reflecting what were viewed as significant achievements during that time. Three of the ten stamps show a science and technology interest and image. In each of these, science and technology is shown in a heroic context. Figure 4.24 shows four stamps from the set. The first stamp defines the set, with the date, the title of the commemoration, a red star and the Soviet crest on a gold background. A hand-drawn firework display (top left) marks the celebratory nature of the set. The remaining nine stamps are embossed with the Soviet Crest and each has a title explaining the image. *Builders of communism* (top right), features a painting by Merpert and Skripkov, with scenes of industrialisation in the background. *Lenin explaining the GÖELRO map* (bottom left) refers to the first unified state long-range plan for the national economic development of the Soviet republic, based on the electrification of the nation. *Dawn of the Five-Year Plan* (bottom right) features a picture of construction work at Romas in 1934. All ten stamps are of the same value, and it is doubtful that anyone other than a collector would see all ten displayed at the one time. All would fall into the classification of providing a lens, encouraging the interested viewer to seek additional information on the historical data presented and informing the viewer of the importance of science and technology in moving Russian aspirations forward.

Figure 4.24: Russia, 1967. *The 50th anniversary of the October Revolution.* **Gibbons catalogue # 3473, 3482, 3475 and 3479.**

Source: Author's collection.

In similar fashion, the USSR 24th Soviet Union Party Congress Resolutions of 1971 were also commemorated with the issue of a set of stamps. Two of the five images relate to science and technology and are shown in Figure 4.25. The congress resolutions refer to Heavy Industry ("Industrial Expansion") and Factory Production Lines ("Increased Productivity"). Appropriate images put the list of resolutions for each industry into perspective. They are low service value for within country use, but they are a lens nevertheless.

Figure 4.25: Russia, 1971. *24th Soviet Union Party CongressResolutions.* **Gibbons catalogue # 3979 and 3978.**

Source: Author's collection.

These series of stamps demonstrate there is a clear desire to educate and inform, exemplified by the statement on the miniature sheet, Figure 4.26, translated as: "the main problem is to provide a significant increase of the material and cultural level of life of the people on the basis of high rates of development of socialist production, an increase in its effectiveness, scientific technical progress, and an acceleration of the growth of the productiveness of labour". A lens, indeed, directed, by its service value, for international circulation, despite the fact that the text is in Russian.

Figure 4.26: Russia, 1971. *24th Soviet Union Party Congress Resolutions.* **Gibbons catalogue #MS 3981.**

Source: Author's collection.

The trend towards representing political anniversaries through a lens of scientific achievement continued until the establishment of the Russian Federation. The USSR issued a long series of issues showing their space achievements, 46 of which featured Yuri Gagarin as the first man in space. The first space research example is a 1951 issue, in which Konstantin Tsiolkovsky (1857–1935), rocket scientist and described as the pioneer of the astronautic theory (Gibbons Russia catalogue # 2012), was featured as one of 16 *Russian scientists* (see Figure 4.16). Tsiolkovsky has been additionally honoured within four other issues. The USSR issued 168 separate sets of space stamps containing approximately 350 different images. One 1961 stamp, issued after the first two manned space flights, is worthy of reproduction. Figure 4.27 looks ahead to *Cosmic flights*, with the images of a space rocket leaving the earth, labeled USSR, and heading into space, shown with representative stars. The images are enhanced by the fact that the stamp is printed upon an aluminium-surfaced paper, emphasising that Russia's space project is distinct and will be utilising new and innovative materials in its research. The small amount of Russian text, the high service value and futuristic theme, in addition to the innovative printing method, suggest that the stamp was produced with international distribution and circulation in mind, this taking place in 1961, when Russia clearly led in the space race against the United States.

Figure 4.27: Russia, 1961. *Cosmic flights*. Gibbons catalogue # 2635.
Source: Author's collection.

The Russian Federation was created in late 1991 and continued promoting space research achievements, producing another 21 sets until 2011, at which time there was a celebration of the 50th anniversary of Yuri Gagarin's first ever manned space flight. An example is shown in Figure 4.28, itself one of a series of space flights to which Russia had invited cosmonauts from other nations in order to elicit additional publicity. This short set of three celebrates the 1978, *Soviet-Polish space flight*, featuring cosmonaut Pyotr Ilyich Klimuk (born 1942), who made three flights into space, and Mirosław Hermaszewski (born 1941), the first Pole in space. The 32k image shows a ship named after cosmonaut Vladimir Mikhaylovich Komarov (1927–1967). A number of Russian cosmonauts were honoured by having research vessels named for them, as well as being celebrated on postage stamps. These stamps, although officially honouring specific cosmonauts, do not show their faces as the dominant image. One has to search the stamp to discover the man who was honoured, which reduces the importance of the foreign cosmonauts in favour of a strong technological message. The political importance of the joint flight is shown, however, by the use of the Polish flag in juxtaposition to the Russian flag either side of the joint mission symbol.

Figure 4.28: Russia, 1978. *Soviet-Polish space flight.* **Gibbons catalogue # 4777–4779.**
Source: Author's collection.

Figure 4.29 differentiates between the number of other science stamps issued by Russia and those that celebrate their space achievements. Following the death of Stalin, science stamps were underrepresented until the event of Sputnik. Similarly, following the breakup of the Soviet Union, science stamps were relatively few. These facts illustrate the importance of science and technology in delivering political messages: during times of upheaval, the messages are less clear.

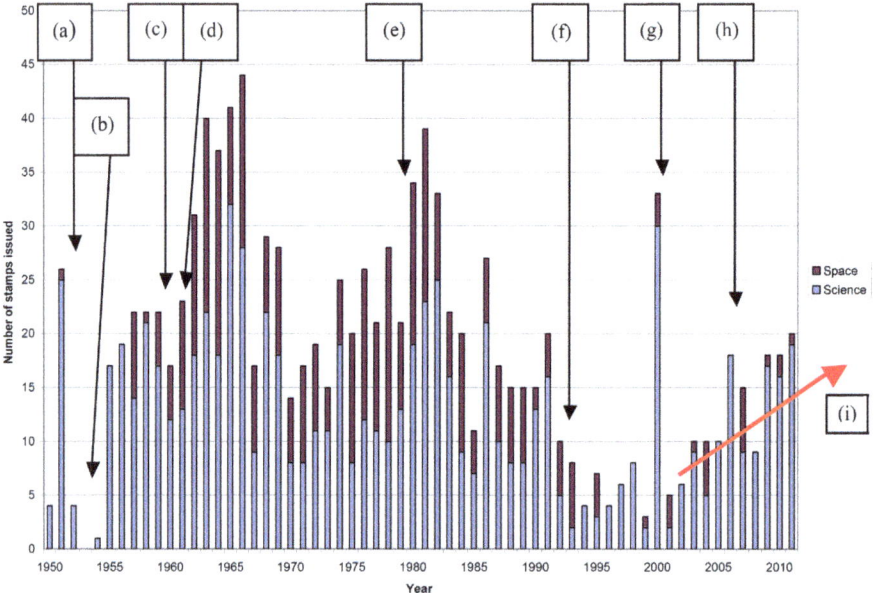

Figure 4.29: Science and space research stamps issued by Russia, 1950–2011.

(a) 1951: The first stamp recognising a rocket scientist, Konstantin Tsiolkovsky (1857–1935), who was eventually featured six times on the stamps of Russia.

(b) 1953: The death of Stalin, with only one science celebration in the next two years.

(c) 1957: The first Sputnik is celebrated.

(d) 1959: The first rockets to the moon.

(e) 1961–1990: The first manned space flight of Yuri Gagarin and intense space research activity, Gherman Titov, the second man in space, Aleksei Leonov, the first space walk, team flights, the establishment of the International Space Station and Russian space flights with cosmonauts of different nations, all celebrated with the most colourful of stamp images, approximately 30% of all science issues, see Table 3.6.

(f) 1991: The breakup of the Soviet Union – relatively few science stamps issued for a few years as the new Russian Federation took stock.

(g) 2000: A peak in issues including two *Millennium sets* of twelve stamps issued celebrating *(2nd issue) science* and *(3rd issue) technology*. The three space research stamps, in a set, describe International Space Cooperation and show the flags of 15 nations, including that of the United States, to demonstrate the level of cooperation.

(h) The period 2001–the present day sees the 50 year anniversaries of the most significant successes of the 1960s.

(i) The red arrow traces the trend of Russian issues. There is again a trend of an increase in the number of science stamps being issued after 2000 which is described in Table 3.5, which we also saw with the stamps of China.

Source: Author's research.

Summary of the science stamps of Russia

Twentieth-century Russia sought to define itself as a modern global power through agricultural reform, rapid industrialisation and the excellence of its science, embracing communist ideals, represented and associated with the powerful icon of the hammer and sickle. Lenin and Marx, founders of the revolution that brought the communists to power, have been celebrated on stamps, and during Stalin's period of power he was the face of the new ideology. Science and technology have subsequently become icons of Russia's progress. Stamps have been used to convey aspirational messages and strongly support Communist Party edicts as artifacts in everyday use. The stamp message exhorts the public to be a part of the revolution as well as announcing a new Russia to the world. The scientist, as an individual, has not been forgotten and, as will be shown in Chapters Five and Six, Russia has celebrated the achievements of more than 250 of its own scientists, approximately 30% of all the scientists identified from the ten countries of my study.

This analysis of Russia's scientific and technological stamps has clearly demonstrated how the stamps were used for political purposes. Many of the images can be classified as lenses in that the message seeks to inform, direct or influence in a positive way the public's commitment to the ideology presided over by the Communist Party. One can distinguish the stamps that are intended for internal use compared to those whose service value implies they will have an international distribution. There is also a trend from images that send an internal message, such as the Five and Seven Year Plans, to one with a broader focus over time as Russia asserts its place in the world. In this respect there are parallels with the changing messages sent by China.

Post-war Germany

A study of German stamps allows me to compare and contrast how the science stamps of East and West Germany tell different messages over a defined period of time. At the end of WWII, Germany was ideologically divided into two distinct countries for 42 years. East Germany, as a member of the Russian dominated Eastern Bloc, issued a series of confirmations of achievements on postage stamps that included strong science messages. These reflect the political emphasis of government at that time. The nature and style of the Russian parent's propaganda is very obvious in its stamp issues over time, as shown in Figures 4.31, 4.33–4.36. Following this review, I discuss comparisons in the issue of science on stamps messages from East and West Germany. Finally, I touch upon examples of direct propaganda stamps in a war situation.

East Germany, the German Democratic Republic, (DDR)

Early stamps from East Germany emphasised that government would pursue an interest in science and technology. Just nine months into the new constitution, in 1950, its first definitive set was issued celebrating the *250th anniversary of the Academy of Sciences,* based in Berlin. Ten stamps were issued, seven of which are reproduced as Figure 4.30.

- 1pf: Leonhard Paul Euler (1707–1783), Swiss mathematician and physicist, who spent a significant part of his scientific life in Berlin, the stamp perhaps claiming him as an adopted son.
- 5pf: Alexander von Humboldt (1769–1859), biogeographer.
- 8pf: Wilhelm von Humboldt (1767–1835), founder of Berlin University.
- 10pf: Hermann Ludwig Ferdinand von Helmholtz (1821–1894), physician and physicist.
- 12pf: Max Planck (1858–1947), physicist, 1918 Nobel Prize in Physics.
- 20pf: Walther Hermann Nernst (1864–1941), 1920 Nobel Prize in Chemistry,
- 24pf: Gottfried Wilhelm Leibniz (1646–1716), philosopher, polymath and mathematician.

The stamps are small, as are many definitive issues. They are printed in a single colour. The portraits are formal and display a gravitas reflecting the status of the Academy of Science. The new country is also, perhaps, staking a claim to these famous celebrities as having been born in what is now East Germany, or, in the case of Euler, having worked there. The set emphasises the scientists from three centuries rather than their scientific achievements, and is a pointer to the policies, under guidance from the USSR, that the new country will follow. The theme is also nation-building, looking back at famous scientists of the past. The message is part mirror and part lens.

Figure 4.30: DDR, 1950. *250th anniversary of the Academy of Sciences.* Gibbons catalogue # (all with prefix E) 20, 21, 23, 24, 25, 27, and 28.
Source: Author's collection.

Just one affirmation of the role of science was made in the 1953 *Five Year Plan*, although the set comprised 18 images of workers fulfilling their roles. A scientist was included shown in his white coat and working with a microscope (Figure 4.31). Other images in the background suggest his work will contribute to industry as a whole. The figure 5, designating the Five Year Plan, is prominent as the reason for the issue. It is a political message. It is a lens exhorting people to support the Five Year Plan and the new reality of the country.

Figure 4.31: DDR, 1953. *A scientist,* a stamp from the definitive set of 18, *Five Year Plan*. Gibbons catalogue # E133.
Source: Author's collection.

In 1956, to celebrate the 110th anniversary of the Carl Ziess Factory, three stamps were issued. Two showed images of the two principal scientists: optician Carl Zeiss (1816–1888), and physicist and optometrist Ernst Karl Abbe (1840–1905). As will be seen in Chapter Five, Carl Ziess is celebrated 13 times on the stamps of East Germany and once by West Germany.

In 1957, East Germany commenced a series celebrating *Scientists' anniversaries*. In the same format as the 1950 set, three anniversaries, each from different centuries, are included within Figure 4.32. Euler is featured again.

- 5pf: Joachim Jungius (1587–1657), mathematician, botanist, logician and philosopher of sciences.
- 10pf: Leonhard Paul Euler (1707–1783), Swiss mathematician and physicist.
- 20pf: Heinrich Rudolf Hertz (1857–1894), physicist/experiments with radio waves.

The set is a continuation of the nation-building theme of the 1950 celebration.

Figure 4.32: DDR, 1957. *Scientists' anniversaries.* Gibbons catalogue # E322-324.
Source: Author's collection.

The tenth anniversary of the country was in 1959. All the designs shown in Figure 4.33 include the East German flag, as well as the images of:

- 5pf: combine harvester.
- 20pf: steel worker.
- 25pf: industrial chemist.
- 50pf: woman tractor-driver.
- 60pf: Ilyushin aircraft.
- 70pf: shipbuilding.
- 1Dm: East Germany's first atomic reactor.

The message is clear: under the protection of the flag, all aspects of life look to science and technological development. Featured are agriculture, manufacturing, research, defence and the provision of infrastructure. Again, these messages are part mirror, part lens.

Figure 4.33: DDR, 1959. *Tenth anniversary of the German Democratic Republic.* **Gibbons catalogue # E455–E464.**
Source: Author's collection.

The East German endorsement of Russian space achievements began in 1959, with the first landing of a Russian rocket on the moon. Two years later, a three stamp set was issued to mark the first manned space flight. Space dominates the publishing programme of the DDR issuing authority, with 260 space related stamps issued over 20 years. Figure 4.34 illustrates the number of space research stamps compared to the total number of science stamps issued by East Germany.

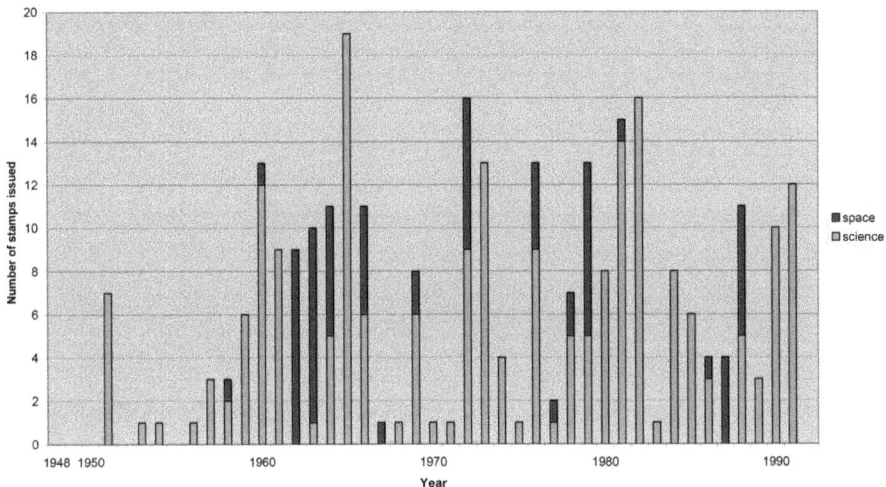

Figure 4.34: Space research stamps issued by East Germany.
Source: Author's research.

This bar-chart illustrates a not dissimilar pattern to that of Russia (Figure 4.29). Science messages endorse public policy from the mid-1950s and both countries feature high levels of activity surrounding space research activities to optimise these successes during the Cold War. Poland, as a satellite of the USSR, exhibits a similar issue pattern.

During 1963, and discussed within the taxonomy results in Chapter Three, Figure 3.32 uses the images of a generic female laboratory worker and a chemical plant within a miniature sheet titled *Chemistry for freedom and socialism*. These two stamps are of a high value, suggesting it was expected that the messages would be carried overseas rather than in country.

In 1964, the DDR, (the acronym was now being used as the main country index), celebrated the 15th anniversary of the country with a 15 stamp set using science and technology as symbols of achievement. Every stamp features the image of an industry and a generic worker carrying his or her tools of the trade. Different background colours are used, with strong colours throughout a consistent design. Female figures are shown in three images.

The designs shown in Figure 4.35 are industrial scenes that illustrate:

- Top row: chemical works, coal surveying, steel mills, recreation.
- 2nd row: culture, shipping, planetarium (optics).
- 3rd row: reconstruction, engineering, agriculture, education.
- 4th Row: new construction, exports, chemical industry, textiles.

Figure 4.35: DDR, 1964. *15th anniversary of the German Democratic Republic*. Gibbons catalogue # E780–E794.
Source: Author's collection.

In the midst of space achievements, the DDR ran a series of four sets that featured *Celebrities' birth anniversaries*, an extension of the earlier issue, shown in Figure 4.32. These issues commonly featured scientists as celebrities. Three of the six stamps from the 1981 fourth series are shown in Figure 4.36 as examples. There are 20 such scientists honoured with stamps following this format issued annually between 1977–1981. They are interesting stamps that show contextually the reason the scientist has been chosen as a celebrity as well as his life dates. The celebrities are:

- 10pf: Heinrich Georg Barkhausen (1881–1956), atomic physicist.
- 25pf: Julius Wilhelm Richard Dedekind (1831–1916), mathematician who did important work in abstract algebra (particularly ring theory).
- 50pf: Adelbert von Chamisso (1781–1838), botanist.

The head dominates the image. The context is shown but, with the exception of the botanist, one would have to be a scientist to determine the significance of the graph shown or the mathematical formula. The three images are good examples of a transition from using the scientist's portrait to tell the message, to the introduction of context.

Figure 4.36: DDR, 1981. The fourth series of *Celebrity birth anniversaries*. Gibbons catalogue # E2316, E2318 and E2320.

Source: Author's collection.

The 35th anniversary of the country was marked with the issue of three stamps solidly representing what the issuing authority saw as two technical achievements (Figure 4.37). A third stamp showed the military and military equipment, which is outside of the scope of my study. The two stamps in Figure 4.37 showcase the development of:

- 10pf: East Ironworks.
- 25pf: Schwedt Petro chemical complex.

These are named on the stamps. This issue is also tied with a new symbol, a tapestry in the national colours and, rather than Russia's iconic hammer and sickle the logo includes a hammer and a pair of compasses. The DDR is unabashedly celebrating what it sees as technology successes, mirrors of achievements.

Figure 4.37: DDR, 1984. *35th anniversary of the German Democratic Republic*. Gibbons catalogue # E2604 and E2606.

Source: Author's collection.

The DDR 1985 set celebrating the end of WWII in 1945 (Figure 4.38) is semiotically integrated. Although different colours, each stamp shows a red flag enclosing a prominent index that defines the years under review, 1945–1985. The title of the stamp issue is shown vertically at right hand side of each stamp. Two icons are used in a combined image: workers, and their tools of the trade. Two images show smiling figures, which are appropriate, while the workers shown engaged in their business are concentrating upon the task in hand. Women are shown on two of the four stamps, indicating the importance of women to the country's technical and scientific advancement. Recognisable people are shown on the lower value stamps. The indices, the country and the value of the service being pre-paid, complete the message. The designs of Figure 4.38 show:

- 10pf: East German cosmonaut Sigmund Jähn and his Russian colleague, Valeri Bykovski, the crew of the Soyuz31/Soyuz29 space mission in 1978.
- 20pf: Adolf Hennecke, a politican, known as an "Activist of the first hour" in East Germany, shown when he was a miner.
- 25pf: Agricultural workers reading a paper.
- 50pf: Laboratory technicians.

Figure 4.38: DDR, 1985. *40th anniversary of the defeat of fascism*. Gibbons catalogue # E2651–E2654.

Source: Author's collection.

The last three science related issues of the DDR, before it was absorbed into the Federal Republic of Germany on 3 October 1990, were:

- 1990, *Lilienthal '91 European Airmail Exhibition*, historic flying machine designs (4).
- 1990, *125th anniversary of the International Telecommunications Union* (4).
- 1990, *21st meeting of the International Astronautics Federation.*

East Germany, the German Democratic Republic (DDR), existed for 42 years. Although in the thrall of Russia, DDR pursued a different regime and style, to a certain extent, in promoting science and technology. The DDR did not promote its ambitions through Five Year Plans. Its successes were celebrated on a time-line of anniversaries of the establishment of the country. The DDR as an entity was celebrated with a strong emphasis on the nation prospering under the national flag. People benefiting from the effects of science and technology was the predominant theme, rather than a textual statement of aspirations. The events that were celebrated had an East German context and did not generally look outside its boundaries, except to mark the Russian space-race triumphs. Its science issuing policy began to look at the outside world in the year before it rejoined with West Germany in 1991.

A comparison of the science stamp issues of East Germany and West Germany

Having studied East Germany it is appropriate to comment upon what policies were adopted by West Germany. During the period 1948–1990, Germany was separated into two countries. Within the boundaries of East Germany, the Western Sector of Berlin functioned as an entity known as West Berlin. Some stamps issued by West Germany were also issued by West Berlin, with the additional index of Berlin added to the issuing authority name. Each regime followed its own political ambitions. East Germany issued approximately twice as many postage stamps as West Germany over the period. Figure 4.39 demonstrates that there were years, occuring approximately seven years apart, when West Germany printed more stamps than East Germany. These were the years in which the (everyday) definitive, multi-value stamps were issued.

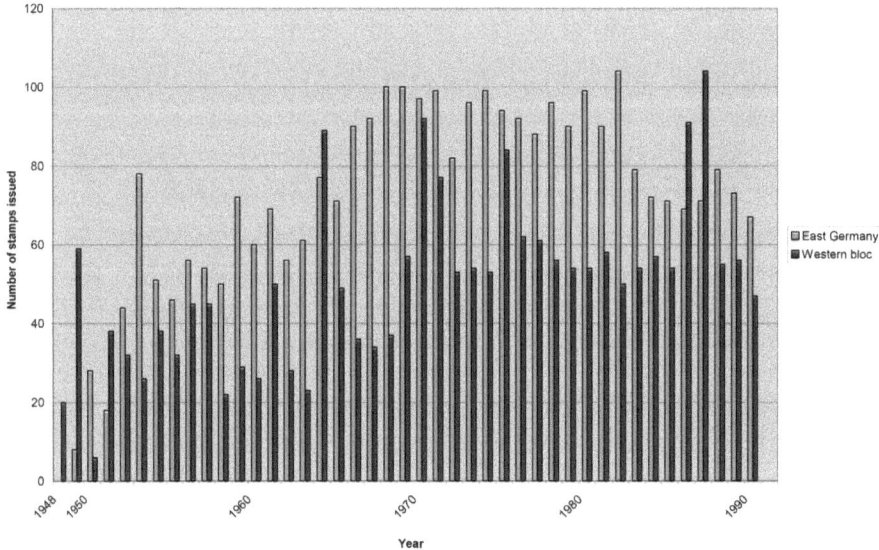

Figure 4.39: Total numbers of stamps issued by the two parts of Germany, 1948–1990.
Source: Author's research.

East Germany under the influence of the Soviet Union issued, in total, 30% more science stamps than West Germany, suggesting that a communist government was more dependent on science for future growth and meeting targets and saw the postage stamp as a more potent communications medium than a democratic government, at least during the 42 years of separation. The total number of all stamps issued during this time was:

- East Germany: 3,061
- West Germany: 1,292
- West Berlin 763:

Geographically, West Berlin was within East Germany. West Berlin issued 65 of its own science stamps, although some issues were identical to the West German issues. In these cases the index, the name of the issuing authority, was changed: "Berlin" has been added as the issuing authority at the bottom right corner of the stamps (see Figure 4.40). Three stamps from the two sets of *Famous German women* of 1986–1989 are shown in Figure 4.40. The celebrants are:

- 60pf : Dorothea Erxleben (1715–1762), the first female medical doctor in Germany.
- 130pf : Lise Meitner (1878–1968), Austrian-born (later Swedish) physicist.
- 140pf : Cécile Vogt (1875–1962), neurophysiologist and pharmacologist.

← 'Berlin' added as the issuing authority

Figure 4.40: West Germany and West Berlin, 1986–1989. The three *Famous German women* images used by both issuing authorities. Gibbons catalogue # 2155, 2159, 2160 and B737, B740 and B741.

Source: Author's collection.

The two parts of Germany handled a specific event differently according to their political imperatives. In the case of *The 100 year anniversary of the telephone*, a user who looks very much like Philipp Reis (1834–1874), is shown on the West German stamp, continuing the promotion of him as the inventor of the telephone. But he is not identified on the Berlin stamp or the East German issue shown in Figure 4.41.

Figure 4.41: West Germany, 1977; West Berlin, 1977; and East Germany, 1976. *100 years of telephone service*. Gibbons catalogue # 1837, B533 and E1833.

Source: Author's collection.

The way that the divided Germanys acknowledged the development of space research is quite different. East Germany, whilst under the influence of the

Eastern Bloc, used the Russian space successes as propaganda for the communist system. 68 space stamps were issued, equating to 24% of all 283 science/scientist stamps issued.

West Germany and the separate West Berlin postal authority did not enthuse over the space-race, and over the 42 years of their existence, as shown in Figure 4.42, they issued only three such stamps, of which two carried no textual explanation of the icon used as the image. The image of the US Space Shuttle, which was being designed in the mid-1970s and made its first orbital flight in 1981, is a recognition of the space programme in the west and was very much in the news when issued, but is treated as matter-of-fact by West Germany. In contrast the 1968 East Germany space stamps were larger than normal, very colourful and often carried textual and additional space motifs to emphasise particular space achievements. An example is shown in Figure 4.43.

Figure 4.42: West Germany, 1975 and 1986. *Industry and technology*, two values from the definitive set of 23, and *Appearance of Halley's Comet*. Gibbons catalogue # 1739, 1743 and 2119.

Source: Author's collection.

Figure 4.43: East Germany, 1978. *Soviet-East German space flight*, one example (from a possible 68 such colourful images). Gibbons catalogue # E2065.

Source: Author's collection.

A total of 8.8% of all the stamps issued by the German postal authorities between 1871 and 2011 have shown a science. The three classifications that dominate these science stamps are (from Tables 3.7 and 3.8):

- A named scientist and an event: 38.2%
- An event or anniversary of science in the abstract: 28.7%
- Scientific images in general: 15.2%

I have discussed above the differences in the stamp issues from East and West Germany during partition. I wondered if, as named scientists are a large component of the total science stamps, the political ideology would differentiate between the science heroes of the east and west, or the country once reunified. The result is shown in Table 4.1.

Table 4.1: The number of stamps of the most celebrated scientists of Germany.

Scientist	Celebration of achievement Germany		
	East	West	New
Carl Zeiss (optics manufacture)	13 x ✓	✓	
von Stephan (Univeral Postal Union)	3 x ✓	2 x ✓	✓
Phillip Reis (telephone)	5 x ✓	2 x ✓	
Max Planck (physics)	3 x ✓	✓	3 x ✓
Johannes Gutenberg (printing)	3 x ✓	✓	✓
Robert Koch (bacteriology)	2 x ✓	2 x ✓	✓
A. von Humbolt (naturalist)	5 x ✓	✓	
Albert Schweitzer (medicine)	4 x ✓	✓	✓

Source: Author's research.

Summary of the science stamps of Germany

From this analysis, it would appear that neither East or West Germany played politics in claiming scientists' achievements for their own ideology during partition, taking into account the fact that East Germany issued 2.6 times as many stamps as the West. The very high number of East German stamps honouring Carl Zeiss is, I believe, related to the fact that the Zeiss Factory, with an international reputation for excellence in optics over a long period, was located in Jena in the East.

Science on stamps: Propaganda in a war situation

From WWII we have examples of specific issues of stamps being used by Germany, (the Third Reich), in conquered countries as propaganda. Included among the wealth of books written about WWII is Albert Moore's extensive analysis of the *Postal Propaganda of the Third Reich* in which he concludes:

> For it is certain that the images and messages seen by Germans on their stamps, postcards and postmarks did have an incredible impact upon shaping public perceptions. Their sheer volume alone undoubtedly made them as important as almost any other medium (Moore, 2003, p. 136).

Moore devotes a chapter of his book to the postal issues of occupied territories. Two interesting examples of how the Nazis used stamps to influence perceptions are shown below. The French *Scientists against cancer,* shown in Figure 4.44, issued in 1941, a year after German occupation, is dramatically different from previous French charity stamps, although it follows an established practice of charging a premium for public health initiatives. The image shows a heroic female, Scientifica, emerging from a clenched fist wielding a sword against a multi-headed snake, the foe cancer. This stamp is a dramatic lens, telling the French people that Nazi Germany is working with them to defeat cancer.

Figure 4.44: France, 1941. *Scientists against cancer/Anti-cancer Fund.* **Gibbons catalogue # 699.**
Source: Author's collection.

Moore also discusses how Joseph Goebbel's Nazi propaganda apparatus sought to celebrate the occupied country's "liberation" by Germany by reverting to German placenames and celebrating local achievements as joint achievements of the locals with Germans. A Polish example is shown in Figure 4.45. Here, the General Government, as the ruling German Military called the Polish captured territory, uses a portrait of Nicholas Copernicus alongside portraits of Germans with an association with the territory. The Polish hero has the same standing as the recognised Germans, implying that is "he was an ideal citizen" (Moore, 2003, p. 87). The celebrants are (left to right): Veit Stoß (1447–1533), German sculptor; Albrecht Dürer (1471–1528), German painter; Johann Christian Schuch (1752–1813), German architect; Josef Elsner (1769–1854), German composer; and Mikołaj Kopernik (1473–1543), Polish astronomer.

4. Stamps as Communicators of Public Policy

Figure 4.45: Poland, 1942. *The third anniversary of German occupation*, celebrating four German artists and Poland's Copernicus. Gibbons catalogue # 451–455.
Source: Author's collection.

This small section illustrates the lengths to which the political perspective can be taken through the medium of science on stamps. It is propaganda.

Having examined how major nations use postage stamps to send messages of scientific progress, I shall now look at the postage stamps of the one continent that did not have an indigenous population when discovered, Antarctica, where science is used as the reason for populating the area.

The Representation of Science and Scientists on Postage Stamps

The stamps of Antarctica as a political medium

"A land of cold truths". Antarctica boasts only explorers and scientists, seals and penguins, ice and snow, ships and flags and whales (Thomas, 2012).

The Antarctic Treaty was constituted after, and as a result of the International Geophysical Year (IGY) of 1957–1958, an international scientific project with 67 participating countries. It marked the end, after Stalin's death, of a long period during the Cold War when scientific interchange between east and west was seriously interrupted. The most dramatic of the new technologies available to the IGY was the rocket. Post-WWII developments in rocketry, for the first time, made the exploration of space a real possibility (National Academy of Sciences, 2005).

The IGY was celebrated on stamps by both the United States and Russia, as shown in Figures 4.46 and 4.47. The Russian stamps show the infrastructures of scientific exploration, *The interior of an observatory* and *A meteor in the sky*, while the third stamp illustrates *A rocket*. For their time, the images are lenses to promote an interest in science.

Figure 4.46: Russia, 1957. *International Geophysical Year.* **Gibbons catalogue # 2093–2095.**
Source: Author's collection.

The United States issue is symbolic invoking Michelangelo's painting *The Creation of Adam*, and an eclipse of the sun, hinting at the creation of the universe. The issue title is quite prominent. This is, again, a lens but less secular than the Russian interpretation of the same theme.

Figure 4.47: United States, 1957. *International Geophysical Year.* Scott catalogue # 1107.

Source: Author's collection.

Poland and East Germany also issued stamps for the IGY, with images of Russian space achievements emphasising the political advantages the Eastern Bloc perceived in these accomplishments. The United States was the only western power to mark the IGY. China was inward looking at the time and did not note the event. Both the Soviet Union and the United States launched artificial satellites for this event. The Soviet Union's *Sputnik 1*, launched on 4 October 1957 was the first successful artificial satellite, Figure 4.48 complements the issues of subsequent satellites shown in this chapter in Figures 4.18 and 4.19. The United States did not celebrate its first satellite on a postage stamp.

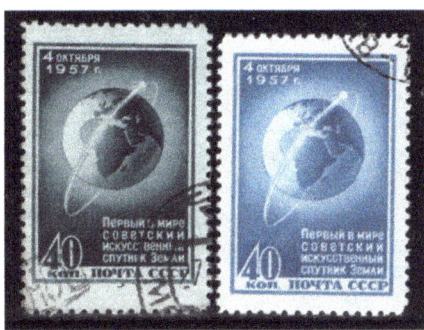

Figure 4.48: Russia, 1957. *Launch of the first artificial satellite.* Gibbons catalogue # 2147 and 2148.

Source: Author's collection.

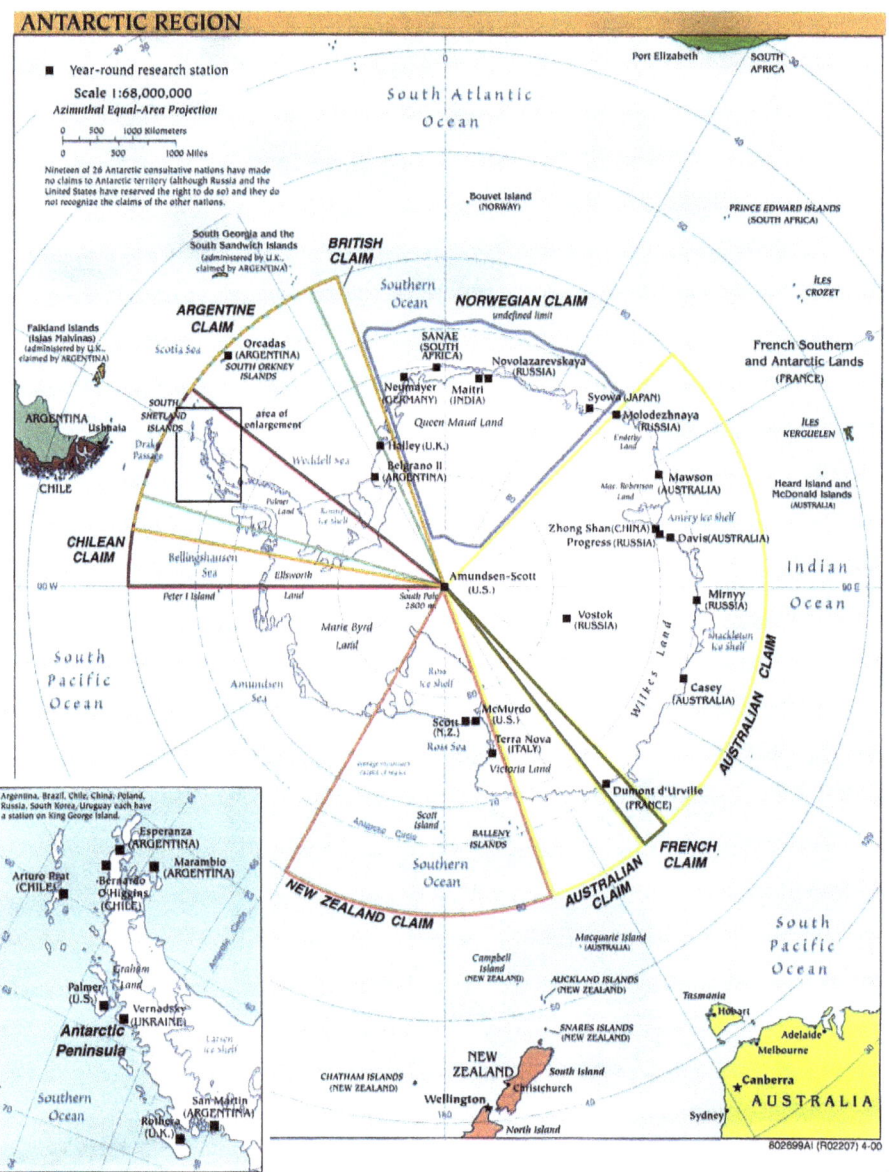

Figure 4.49: Map of Antarctica showing the Antarctic Treaty divisions.
Source: Wikipedia Commons.

I have studied the stamps from the Australian, New Zealand, English and French Antarctic Territories to test Altman's assertion:

> A peaceful example of the use of stamps as part of disputes over sovereignty is from the Antarctic, where there is no full international recognition of the claims of various countries to parts of the continent ... in the absence of international recognition, the creation of special stamp-issuing authorities represents one of the ways in which the various contending powers seek to assert their claims to sovereignty (Altman, 1991, p. 23-24).

These are four territories of countries in this study that abide by the terms of the Antarctic Treaty, which came into force on 23 June 1961. The treaty was ratified by the 12 countries then active in Antarctic science, pledging peaceful uses of the continent and scientific cooperation. These countries maintain research establishments and overtly cooperate with other countries in their research activities. The other three countries who claim Antarctic Territories are Norway, Argentina and Chile, as is shown in Figure 4.49. The United States bases its Antarctic Research at McMurdo Sound within the New Zealand Ross Dependency. Logically, therefore, any stamps have to reflect research activity. Do they constitute unique areas of science related endeavour? Do the stamp images that are used in these territories, which have issued separate stamp issues since the 1950s, fall into the category of science? They nearly all do, except that images of the flora and fauna of the territories are used on as many as half of the stamps issued. The other half show science or scientists. As might be expected, there are a few territory stamps that reflect what are considered significant events of the home country. One example is the birth bicentenary of Michael Faraday, with a set of stamps commemorating explorer Sir James Clark Ross, overprinted with a legend in blue, "200th anniversary M Faraday 1791–1867". One symbol is common to many of the Antarctic stamps, a map of the area.

The heroes of Antarctic exploration

Each Antarctic territory celebrates its own heroes of the unique science associated with exploration and research in the adverse conditions of the Antarctic. As Moyal puts it: "survey expeditions [were deployed] to remote regions of the world for the purpose of mapping and charting their unknown shorelines, opening safe passages for trade routes and consolidating Imperial influence. Linking geography and science these surveys adopted increasingly sophisticated knowledge of astronomy, meteorology and magnetism with navigational skills" (Moyal, 2012, p.8).

These survey expeditions were prompted by the political ambitions of European countries and resulted in the discovery of Australasia, the Southern Islands of the Atlantic and Pacific, and Antarctica itself. These discoveries prompted the epic research endeavours of those Iain McCalman has christened as "Darwin's Armada", culminating in the theory of evolution proposed by Darwin and Wallace (McCalman, 2009). The heroic age of Antarctic exploration evolved at the beginning of the twentieth century and is well recorded in the historical and image examples by the postal authorities of Antarctica with the appropriate recognition of the explorer's influence, irrespective of nationality.

A number of the first stamps published by the Antarctic Territory postal authorities are worthy of detailed examination.

Australian Antarctic Territory (AAT)

According to Burke's *The Stamp of Australia*: "The first Australian Antarctic Territory stamps were issued in 1957 under Collas (Phil Collas, Philatelic Officer in the Post Master-General's Department 1951–1969). Collas was a fervent believer that one of a country's most effective means of establishing sovereignty over questionable territory could be achieved through the issue of distinctive stamps" (Burke, 2009, p. 199).

Figure 4.50 illustrates a 2001 stamp set from the AAT. As is common with Antarctic issues, it includes a map of Antarctica and shows the location from which the main image is taken.

The viewer is expected to recognise that the research activities at each research station are much broader than the specific (mirror) examples shown. As will be seen, with the territory stamps, the local fauna provide an identifier to location. The set is tied together by the full name of the authority, which is consistently shown at right angles to the basic stamp design, the appearance of the map of Antarctica, a white arc of a circle containing an icon of the area, and a scenic background. Each stamp shows one of the four permanent research stations on the map, named for an explorer or Australian Governor:

- Davis Base was named after Captain John King Davis (1884–1967), a famous Antarctic navigator and captain of Mawson's ship, and a member of the Australian National Antarctic Research Expedition Planning Committee until his death in 1967. Atmospheric science, a part of research that takes place on this base, is used as the basis for an image.
- Casey Base was named for Governor General Casey (1890–1976), the 16th Governor General of Australia. The image shows an interest in diatom research.

- Macquarie Island, discovered in 1810, is named after Lachlan Macquarie (1762–1824), the then governor of New South Wales.
- Mawson Base is named for Sir Douglas Mawson (1882–1957), the famous Antarctic explorer.

Figure 4.50: Australian Antarctic Territory, 2001. *Antarctic research*. Renniks catalogue # AAT160–AAT164.
Source: Author's collection.

Figure 4.51 uses five stamps of a 1973 definitive set to tell a story. The story is part of the *Antarctic food chain*. The stamps show:

- 1c: plankton/red shrimp.
- 7c: Adelie penguins.
- 9c: leopard seal.
- 10c: killer whale.
- $1: sperm whale and squid.

Together, the stamps illustrate the reliance of each species upon the others. These five stamps are a lens of enquiry into the relationship of species in Antarctica. Six of the other stamps in the set, discontinuous in design, show various explorers' airplanes and their mode of travel.

Figure 4.51: Australian Antarctic Territory, 1973. *AAT food chain*. Renniks catalogue # AAT023, AAT025, AAT027, AAT028 and AAT034.

Source: Author's collection.

The New Zealand Antarctic Territory, known as the Ross Dependency

Figure 4.52 is from the first Ross Dependency issue of definitive stamps. The stamp, issued for use by members of the New Zealand Antarctic Contingent in 1957, is worth study. It is included in my study as a science stamp, as I have determined that the early explorers, as previously discussed, had a scientific ambition within their exploration objectives. Semiotically, it can be deduced that thought has gone into the design. The stamp is printed in imperial red, as was most of the globe during the glory days of empire. The issuing authority is clearly stated as Ross Dependency; no mention is made of New Zealand. The currency (the index) is given as New Zealand pound sterling (decimal currency was adopted in 1967 and this definitive stamp was re-issued). The map of Antarctica shows the land claimed as the Ross Dependency. Two key figures of the heroic age of Antarctic exploration are shown, Sir Ernest Shackleton, and Captain Robert Falcon Scott, from the British Antarctic Expedition of 1910–1913. The two iconic images are enveloped with (the victors') laurel garlands.

A second stamp in the set shows the Royal Navy ship HMS Erebus against an Antarctic background. HMS Erebus was one of the ships of Sir James Clark Ross when he discovered McMurdo Sound in 1841.

Figure 4.52: Ross Dependency, 1957. One of four *definitive stamps*. Campbell Paterson catalogue # RD5.

Source: Author's collection.

The British Antarctic Territory (BAT)

The stamps of BAT tend to follow the heroic age of Antarctic discovery format and image. One such stamp, from a set of four, is shown as Figure 4.53. Captain Scott is shown in Naval dress uniform in 1904 on the stamp, with a drawing of a sled being pulled by man-power as occurred on the 1911–1912 expedition to the South Pole.

Figure 4.53: British Antarctic Territory, 1987. *75th anniversary of Captain Scott's arrival at South Pole.* **Gibbons catalogue # 155.**
Source: Author's collection.

The French Antarctic Territory

Issues from the French Antarctic Territory include the stamp shown in Figure 4.54. The symbolism used here is quite complicated. The theme is declared to be "1968—Vingt ans d'activités of expeditions polaires Françaises". Shown against an Antarctic landscape containing human figures are three vehicles used in exploration: a tracked tractor pulling a sled, a helicopter, and an airplane. At the top left of the stamp there is a symbol, perhaps a globe through which is a pair of signaling flags. I have not been able to decipher this sign, nor can I explain the row of Mediterranean homes beneath the text at the bottom left of the design. My confusion makes this a lens for me.

Figure 4.54: France, 1968. *20 years of Polar expeditions*. Gibbons catalogue # 1806.
Source: Author's collection.

The celebration of Antarctic scientific research by countries outside of those claiming territorial rights

Germany's continuing interest in Antarctica is reflected through a number of stamp issues. Johann Georg Adam Forster (1754–1794), was a German naturalist and ethnologist who travelled on several scientific expeditions, including James Cook's second voyage to the Pacific. He is commemorated with an image of the Antarctic Research Station named after him (Figure 4.55). It is the establishment of the research station rather than the man being commemorated here, although in using Forster's name the country has acknowledged his achievements. The stamp is a mirror, it states that East Germany has a research station in Antarctica.

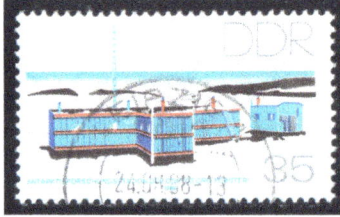

Figure 4.55: DDR, 1988. *12th anniversary of Forster Antarctic Research Station,* named after Georg Forster. Gibbons catalogue # E2863.
Source: Author's collection.

West German interest in Anarctica is centred on the Von Neumayer Research Station in the Weddell Sea. Georg von Neumayer (1826–1909) had established the German Commission for South Polar Exploration, which culminated in the First German Antarctica Expedition in 1901, the so-called Gauss expedition. Two stamps are shown in Figure 4.56. The first image is modern and shows a gloved and hooded surveyor, his equipment, a tracked vehicle, and a group of huts comprising the research station against an Antarctic mountainous background. The sun is high but is showing a white light. The second stamp, issued in the same year, celebrates the 20th anniversary of the Antarctic Treaty,(Antarktis-Vertrag), and includes the dates 1961–1981. The image is a 3-D representation of the familiar map of Antarctica.

Figure 4.56: West Germany, 1981. *Georg Balthazar von Neumayer, polar explorer and scientist, polar research* **and** *20th anniversary of the Antarctic Treaty.* **Gibbons catalogue # 1964 and 1981.**
Source: Author's collection.

On 3 October 1990, the territory of the Democratic Republic, East Germany, was absorbed into the Federal Republic of Germany. The first two stamps issued by the reunified country showed the national colours—black, red and yellow—and followed the theme of unity. Antarctic research was not forgotten and, in 2001, a miniature sheet was issued to mark a centenary of German Antarctic research (Figure 4.57). The name of Von Neumayer and his life-dates are shown. Von Neumayer's original ship, the survey barquentine Gauss represents the original exploration vessel, whose course is shown on the global map. The Polarstern, a modern exploration ship, brings the navigation interest up-to-date. A huge ice-flow that provides a visual context can only be seen if the miniature sheet is viewed in its entirety. The message is one of continuity.

The Henryk Arctowski Polish Antarctic Station is a research station on King George Island, off the coast of Antarctica, named for Henryk Arctowski (1871–1953), a meteorologist on the Belgica expedition of 1897-1899. Figure 4.58 includes a 1973 celebration of eight Polish explorers, including Arctowski ,who is shown on a medallion against an Antarctic view with Adelie penguins. The middle stamp in the figure is one of six Polish scientific expeditions. The main image is of seals on ice floes off-shore from the Arctowski Antarctic Station. The Antarctica map shows the location of King George Island and, at the other extreme, the location of the 1959 expedition. The third stamp shows the research station under a sky lit by the aurora australis. The Polish flag is flying above the huts and a helicopter is parked close to two scientists working from a tent in the left foreground. The stamps mirror the reality of how Poland sees its interest in the area.

Figure 4.57: Germany, 2001. *Centenary German Antarctic Research*. Gibbons catalogue # MS3084.

Source: Author's collection.

Figure 4.58: Poland, 1973, 1980 and 1982. *Henryk Arctowski (1871–1953), meteorologist and explorer*; *Polish (1959) Antarctic Scientific Expedition*; and *50th anniversary of Polish Polar Research*. Gibbons catalogue # 2267, 2673 and 2845.

Source: Author's collection.

4. Stamps as Communicators of Public Policy

Russia has contributed approximately one-third of the stamps I have examined during my study. That proportion is substantiated within this section looking at the stamps celebrating Antarctic research.

Russia has maintained a sustained interest in research in Antarctica since 1820 that is represented by messages carried on its postage stamps. Two Russian sailors have here been judged to be heroes of science, by virtue of the number of times that they have appeared on postage stamps. They are Mikhail Lazarev (1788–1851), fleet commander and explorer, and Fabian von Bellinghausen (1778–1852), naval officer, who are celebrated as leaders of the second Russian circumnavigation of the globe. Figure 4.59 shows the 1950 two stamp issue to mark the *130th anniversary of Russia's first Antarctic Expedition*. Stamp one shows and names both Lazarev and Bellhausen. A globe illustrates Antarctica and the voyage route in relation to Europe and Africa. The background contextually includes a polar scene with penguins. The second stamp has images of the ships Mirnyi and Vostok and a more detailed route map of their exploration of Antarctica. Again, the ubiquitous penguin is shown, along with an albatross in flight. Both stamps are inscribed with 1820, the year of the Russian expedition's discovery of the Antarctic mainland. The message is a straight forward commemoration of the achievement. Lazarev and von Bellinghausen also feature on five other Russian stamps.

Figure 4.59: Russia, 1950. *130th anniversary of first Antarctic expedition*, featuring the images of explorers von Bellinghausen and Lazarev. Gibbons catalogue # 1647–1648.
Source: Author's collection.

The first Russian Scientific Antarctic Expedition was led by Mikhail Somov, and took place over 1955–1957. An objective of the expedition was to establish the Mirnyi research base. In the first first stamp in Figure 4.60, the route taken by the supply ship Lena is shown, as is the ship itself, a badge of the expedition and again shows penguins for additional context. Somov has also featured on later stamps, two (of another five) of which are also shown in the figure.

The miniature sheet illustrates the Antarctic drift of the research vessel Mikhail Somov and the appropriate map. The third image commends Somov as a polar explorer.

Figure 4.60: Russia, 1956, 1986 and 2000. *Soviet scientific Antarctic expedition*; *The Antarctic drift of the Mikhail Somov*; and *Somov the polar explorer.* **Gibbons catalogue # 2026, MS 5695 and 6890.**
Source: Author's collection.

The tenth anniversary of the Soviet Antarctic Expedition was celebrated in postal style with three triangular stamps printed se-tenant to form a larger triangle (Figure 4.61). The three main images follow the pattern of symbols used in Russian Antarctic issues. The map of Antarctica extends beyond the triangular stamp and has a label, in essence an extension of the stamp making it diamond shape but of no value in postal terms. The whole map repeats and names the main areas of Soviet exploration and the ten year motif is strong. The two other values are images of the supply ship Ob in the pack ice, attended by a airplane and a snow tractor. The interested observers of this scene include humans and four penguins. The final stamp shows two snow tractors and a communications installation.

Figure 4.61: Russia, 1966. *Tenth anniversary of the Soviet Antarctic expedition*. Gibbons catalogue # 3251–3253.

Source: Author's collection.

Figure 4.62 shows two stamps that celebrate the *150th anniversary of Bellingshausen and Lazarev's 1802 Antarctic expedition*. The second image is diverse and complicated and features a modern research station.

Figure 4.62: Russia, 1970. *150th anniversary of Bellingshausen and Lazarev's Antarctic expedition*. Gibbons catalogue # 3788–3789.

Source: Author's collection.

The enlargement, shown in Figure 4.63, can be seen to includesa space rocket after take off, although the rocket appears tethered. A stratospheric or a weather balloon is also shown. There is also a representation of the irregular radiating wave circles of the Doppler effect. The huts within the station are festooned with communications equipment, as is a snow plough. All of this suggests that the latest technology is being used to advance the research being undertaken.

The Representation of Science and Scientists on Postage Stamps

Figure 4.63: Enlargement of the second stamp from Figure 4.62. The symbols indicate high-technology involvement in the Russian research.
Source: Author's collection.

The final Russian set I shall review in this section is shown in Figure 4.64. It celebrates the *50th anniversary of Antarctic research*. The set is tied together through the rich blue background and the consistent placement of the Antarctica map on each of the three stamps. The effect is almost like looking out of a cave into daylight. The stamps show, from left to right: the diesel electric ship *Ob*, this time named as an icebreaker rather than as a supply ship off the Antarctic coast and the Mirnyi Research Station under the Russian flag; an *IL-76TD* airplane and 717 scientific ship *Academician Fedorov*; under-ice research and sledge transport train with the Russian Federation flag.

Figure 4.64: Russian Federation, 2006. *50th anniversary of Antarctic research***. Gibbons catalogue # 7387–7389.**
Source: Author's collection.

China highlighted its own Antarctic research interests in 2002. The set of three shown in Figure 4.65 is representative of stamps issued with this theme, featuring unique fauna, the aurora australis, and an Antarctic scene.

4. Stamps as Communicators of Public Policy

Figure 4.65: China, 2002. *Antarctica research*. Gibbons catalogue # 4736–4738.

Source: Author's collection.

China's involvement in the region, in which two of its three permanent research stations are based within the Australian Antarctic Territory, was raised by the Fairfax Media in January 2010: "China's interest in Antarctica has developed rapidly in the past decade. Two of its three bases are in the Australian Antarctic Territory, and there is a record of growing co-operation on scientific research" (Darby, 2010).

The Antarctic Treaty

The Antarctic Treaty is regularly celebrated in its own right by both the Antarctic territories and their parent countries. The Territory map provides the context for many issues.

Figure 4.66 illustrates two aspects of freezing. The 6 cent stamp shows Sastrugi ice and the 30 cent stamp shows sea pancake ice. The dates that the Treaty had run, 1961–1971, are shown textually on both stamps in addition to the territory name.

Figure 4.66: Australian Antarctic Territory, 1971. *10th anniversary of the Antarctic Treaty*. Renniks catalogue # AAT019–AAT020.

Source: Author's collection.

Figure 4.67 tells its story through the representation of an unnamed scientist. It celebrates the *12th Antarctic Treaty Consultative Meeting, Canberra, 1983*. The image on the stamp shows an expedition member with the background featuring a list of the scientific disciplines being pursued on the continent. The text is explicit in explaining the reason for the forthcoming meeting.

Figure 4.67: Australian Antarctic Territory, 1983. *12th Antarctic Treaty Consultative Meeting, Canberra1983*. **Renniks catalogue # AAT061.**
Source: Author's collection.

The various anniversaries of the Antarctic Territory have also been celebrated by Russia and the US, with the dominant motif being the Antarctica map, with which we are familiar.

The Ross Dependency set (Figure 4.68) is tied by a continuous mountain range seen in relief, and a spectrum that runs through the five stamps, which are likely to be viewed as a set by stamp collectors. Explanatory text is included for each value printed in a colour from that part of the spectrum shown on that stamp. The 50 cent stamp has a map showing the part of the continent claimed by New Zealand as the Ross Dependency. The $1 stamp shows penguins, an adult and a chick, and the descriptive text "Antarctica shall be used for peaceful purposes only". The $1.80 stamp has the image of a surveyor and his equipment, including a communications transmitter. The additional text, from the Treaty, reads: "Freedom of scientific investigation in Antarctica". The image for the $2.30 issue shows five signal flags being blown in the wind and text: "International cooperation in scientific investigation in Antarctica". The final stamp, $2.80 in value, illustrates a Weddell seal, with text reading: "Preservation and conservation of living resources in Antarctica". The set of stamps also shows a silver fern in the top left corner of every stamp in order to associate the Dependency with New Zealand.

Figure 4.68: Ross Dependency, 2009. *50th anniversary of the Antarctic Territory*. Campbell Paterson catalogue # RD113–RD117.

Source: Author's collection.

Figures 4.69 and 4.70 show the 2009 *50th anniversary of the Antarctic Territory* for the French Antarctic Territory and Russia, whose messages follow a familiar format developed by the issuing authorities.

Figure 4.69: French Antarctic Territory, 2009. *50th anniversary of the Antarctic Territory*. WNS catalogue # TF039.09.

Source: Author's collection.

Figure 4.70: Russia, 2009. *50th anniversary of the Antarctic Treaty*. WNS catalogue # RU085.09.

Source: Author's collection.

So it would appear that political machinations are being played out, even today, with postage stamps as the communications medium.

Antarctic unique designs and messages

Figures 4.71–4.74 illustrate the progression of images of four heroes of Antarctic science, from sketch to portrait to the modern use of photographs now commonly used. In each case the figure starts with a simple sketch showing the named explorer in their favoured protected clothing, through a contextual examination of their exploration, to a later record shown by the iconic photographs taken at the time, except in the case of Amundsen, whose wearing of a bowler hat, suit and tie appear incongruous.

Raoul Amundsen (1872–1928)

Figure 4.71: Ross Dependency, 1995. *Antarctic explorers*; Russia, 1972 and 2011. *Birth centenary* and *The race to the Pole*, one of a set of five; French Antarctic Territory, 2012. *Commemoration*. Campbell Paterson catalogue # RD37 and RD123, Gibbons catalogue # 4079.

Source: Author's collection.

Captain Robert Falcon Scott (1868–1912)

Figure 4.72: Ross Dependency, 1995 and 2011. *Antarctic explorers* and *The race to the Pole*; British Antarctic Territory, 2008 and 2012, *Explorers (and their ships)*, and *Centenary of the British Antarctic Expedition 1910–1913*. Campbell Paterson catalogue # RD38 and RD125. Gibbons catalogue # 468. The final picture of the four is from a set of 16 of Herbert Ponting's iconic photographs.

Source: Author's collection.

Sir Ernest Shackleton (1874-1922)

Figure 4.73: Ross Dependency, 1995. *Antarctic explorers*; Ireland, 2001. *Celebrating the Millennium*, one of a set and miniature sheet of six; British Antarctic Territory, 2008. *Explorers (and their ships)*; Ireland, 2004. *90th anniversary of Shackleton's Antarctic expedition*, set of four. Campbell Paterson catalogue # RD39. Hibernian catalogue # C1065. Gibbons catalogue # 469. Hibernian catalogue # C1274-1277.

Source: Author's collection.

Richard E. Byrd (1888-1957)

Figure 4.74 shows two US issues that celebrate Richard E. Byrd. The middle stamp, issued by the Ross Dependency, names Byrd and his colleague Floyd Bennett (1890–1928), an American aviator who piloted Byrd on his attempt to reach the North Pole in 1926. The image of an airplane on skids, rather than wheels, for Antarctic take-off and landings is included. The United States 1933 issue, announces the two-year expedition by Admiral Byrd, largely self-financed in the middle of the Great Depression. Antarctica, at the time, was of importance to the US and it is noted that Byrd's mission is to return to Little America, named in 1929 and marked on the map on the 3 cent stamp. Also shown is 1926, the date of the first Atlantic flight. The third stamp shows Byrd's name, image and a map showing the limited scope of the expedition's incursion into Antarctica. The United States makes no territorial claim upon the Antarctica and undertakes its research through the Ross Dependencies, McMurdo Station.

Figure 4.74: United States, 1933. *Byrd Antarctic expedition*; Ross Dependency, 1995. *Antarctic explorers*; United States, 1988. *Antarctic explorers*. Scott catalogue # 733. Campbell Paterson catalogue # RD40. Scott catalogue # 2388.

Source: Author's collection.

Of the ten countries chosen for this study, it would appear that all have an interest in Antarctica. The principal characters of the heroic age of Antarctic exploration are celebrated across the postal authorities, although each country, in accord with their stamp selection processes, has also celebrated their native sons. Although they do not claim a part of Antarctica in their own right, Germany, Poland, Russia, the United States and China record their interest in the area through a celebration of the Antarctic Treaty, their maintenance of research stations and acknowledgement of contributions to the opening of the Territory. Altman's claims are justified, it would seem, although limited in their scope for, as we have seen, all authorities advertise their commitment to the ideals of the Treaty.

Further observations concerning stamps as communicators of public policy

This book questions how postage stamps might influence the public awareness of science. The mirror or lens question is an additional pointer to help with answers. I have used it in my analysis of the intent of the message the stamp is delivering. Does the stamp's message reflect the public awareness of the science at the date of issue, or is the message intended to challenge, to be controversial? The definition of lens suggests that a lens message is one that is seeking a change of behaviour by the viewer, the interpreter of the message. Many public health issues and public policy issues are included as lenses and it has to be said that some of the most forceful and memorable messages fall into this category. They have also meant a most careful evaluation of whether they are talking about a science. The New Zealand stamps reproduced in Figure 4.75 dramatically raises the issue of the country's political decision to be anti-nuclear and to refuse entry of nuclear powered ships into New Zealand waters. The first stamp features the international peace symbol superimposed upon an innocent face and uses a view of the earth taken from space as an eye in order to emphasise the message, but it is not a science stamp. The value at which this stamp is issued signals that the message was intended to be seen internationally. It is a powerful message, and no text is deemed necessary. The second stamp, titled *Leading the Way – Nuclear Free 1987*, is from the New Zealand Millennium series. The stamp is symbolically rich, showing a nuclear protest march, a background with a newspaper headlining the denial of US nuclear-powered ships into New Zealand waters, a New Zealand ship and the New Zealand flag. This is an extremely political stamp making a statement on world-politics.

Figure 4.75: New Zealand, 1995 and 1999. *Nuclear free* and *Millennium issue*. Campbell Paterson catalogue # S516 and SH108.
Source: Author's collection.

The Royal Mail guidelines state that postal commemorations should "celebrate the British contribution to the world and reflect the many and varied aspects of the British way of life", (Royal Mail, 2012). Assuming this objective will be common to most postal authorities, it is hardly surprising that good news is commemorated rather than the bad. Affirmation of positive achievement is the normal message. I reproduce an image below (Figure 4.76) in which the message is not good. Russia issued the stamp in 1991 on the *Fifth anniversary of the Chernobyl Power Station disaster*, which occurred in the Ukraine on 26 April 1986. This date is shown in text on the image. The image is described as "radioactive particles killing vegetation". It is certainly a lens, reminding the world of the effects of such an accident.

Figure 4.76: Russia, 1991. *Fifth anniversary of the Chernobyl Power Station disaster*. Gibbons catalogue # 6221.
Source: Author's collection.

Medical research and practice are definitely forms of science, and I have discussed in Chapter Three my inclusion of medicine and public health in this study. Epidemiological concerns have been brought to public attention with a range of interpretations illustrating the messages conveyed on stamps. The underlying messages are about prevention and medical science supporting the public good.

I shall now comment upon how countries have approached the awareness of cancer, mainly breast cancer, as an aspect of public policy. I have chosen this campaign as it covers the best cross-section of the countries involved in the study. Only New Zealand and Great Britain have never issued a stamp to focus upon a public health issue. New Zealand tends to publish positive images of itself, almost as if issues are seen as tourism advertisements. Great Britain has focused upon medical achievements featuring the scientist, achievement and solutions, without drawing attention to the problems being addressed. The other two countries who are not represented in Figure 4.77 are Ireland, who in 1978 highlighted the *Eradication of smallpox* in that country, and Russia, whose two postal campaigns have been a 1993 *AIDS awareness* and a 1995 *Anti-drugs campaign*. The United States is the country who has used stamps to publicise health campaigns the most frequently. The US Postal Service raised an early awareness issue via its 1965 *Crusade against cancer*, (the stamp I have used elsewhere in order to illustrate my definition of science on stamps: Figure 3.1). Other campaigns include 1971 *Drugs*, 1993 *AIDS*, 1999 *Prostate cancer*, 2001 *Diabetes*, 2005 *Child health* and 2008 *Alzheimers awareness*.

The images used are mostly obvious, but it is worth noting that the technology illustrated upon the West German *Cancer prevention through medical check-ups* issue (20f value) is a scintigram, showing the distribution of a radioactive isotope in the body as an aid to diagnosis; the second Chinese anti-cancer campaign, shows diagnosis by thermography. These stamps are lenses, invoking a change of behaviour by the general public through awareness of the campaigns.

4. Stamps as Communicators of Public Policy

Figure 4.77: Australia, 1997. *Breast cancer awareness*; France, 2005. *Breast cancer awareness*; West Germany, 1981. *Cancer prevention through medical check-ups*; Germany, 2001. *Health awareness issues, (cancer)*; Poland, 2002. *Fight against cancer;* China, 1989. *Anti-cancer campaign*; United States, 1996. *Breast cancer awareness*.

Source: Author's collection.

Postage stamps have been used as a medium of charity collection for many years, helping to support victims of disasters and public health issues.

The Breast Cancer Research semipostal stamp shown in Figure 4.78 was issued on 29 July 1998, at a first day ceremony held at the White House. It was the first semipostal in US history. As of October, 2012, the stamp has raised over US$76.3 million for breast cancer research. By law, 70% of the net amount raised is given to the National Institutes of Health, and 30% is given to the Medical Research Program at the Department of Defence. Designed by Ethel Kessler MD of Bethesda, the stamp features the phrases "fund the fight" and "find a cure", along with an illustration of a mythical goddess of the hunt by Whitney Sherman of Baltimore. This stamp is assuredly a lens to have invoked such a large monetary response.

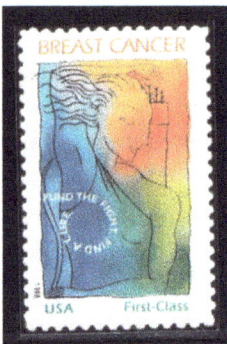

Figure 4.78: United States, 1998. *Breast Cancer Research Stamp*. Scott catalogue # SP 1.
Source: Author's collection.

Stamps as communicators of public policy

The examples shown in this chapter have shown that political messages are sent via the mail. The countries whose stamps have been described are from those countries going through the greatest political and constitutional change during a period when stamps moved from being solely a fiscal receipt for service to a communications medium: China, Russia and the Germanys—the one-party constitutions. The use of science on stamps to promote public policy is more subtle in democracies. Australia, Great Britain and the United States use history and most notably anniversaries as the reason for scientific celebration. Uniquely, space fantasy has been a common theme, allowing countries not directly or strategically involved in space research to imply their interest in what might happen. The older, more established democracies have written into their stamp issuing policy, as far as can be determined, that the images on postal stamps should not be used for the promotion of politics and religion. I did ask the archivist of Australia Post if he was aware of any instance that government had requested a specific stamp issue in the last 20–30 years. He was adamant that that his records contained no such instance. There are instances when countries such as New Zealand do ignore the religious stricture, as do, seemingly, all the western countries which publish Christmas stamps and celebrate other religious festivals. Science, however, has not been used within a religious theme.

When a country adopts a policy that emphasises science and technology as a major platform of development, it is logical that this message be communicated via all the media available to it, including stamps. All countries, in promoting their historical successes and their influences on the world, have celebrated science achievement on postage stamps.

5. On Being First

> The postage stamp is a unique kind of sign, with an impressive capacity to convey a number of messages in a very confined space (Child, 2008).

In the previous chapter I have been able to review the use of stamps to deliver political messages. In this chapter, I look at how the representation of scientists on stamps has developed over time and investigate examples of stamps being used as the vehicle to substantiate 'firsts' in science.

On being first

Robert Merton's "Priorities in Scientific Discovery" (1957) discusses the way that science is structured, noting the importance attached to the date of a discovery and the world's acknowledgement of the achievement. Disputes are commonplace regarding such recognition, and Merton states that it is more likely that the case will be argued by the institution than the scientist. The scientist will often accept that science is a developing understanding of the world and that many research projects are conducted in parallel. The establishment, here in the form of postal authorities, in some cases as an agency of government, enters the fray of controversy in the very public arena of stamp issue. Merton's comments on being first are reiterated by Collins and Pinch:

> Moreover, the mass media may be used by scientists as platforms to assure their priority in discovery – a well known phenomenon in the sociology of science (Collins and Pinch, 1998, p. 142).

Within the context of this study, I anticipate that it is a national institution, the postal authority, representing the aims and ambitions of the state, that chooses to honour specific scientists. As Merton describes it:

> In a world made up of national states, each with its own share of ethnocentrism, the new discovery rebounds to the credit of the discoverer not as an individual only, but also as a national of the state (Merton, 1957, p. 641).

Merton argues that this is a function of the way that science is organised. Conflicts over who was first to make a scientific discovery are more than likely to be pursued at the institutional rather than personal level. (Merton does, however, illustrate his paper with several examples of infamous arguments, such as those between Galileo and four would-be rivals, between Isaac Newton and Robert Hooke, and Michael Faraday's conflict with his one-time mentor,

Sir Humphry Davy). As a contrast, Merton also acknowledges the well known courtesies that potential rivals for the title of first, such as Charles Darwin and Alfred Russel Wallace, as the discoverers of the rule of natural selection, modestly extended to one another. He has chosen not to emphasise his list of priority fights explaining:

> It is enough to note that these controversies, far from being a rare exception in science, have long been frequent, harsh, and ugly. They have practically become an integral part of the social relations between scientists (Merton, 1957, pp. 636–637).

Merton suggests that it is "institutions rather than individual scientists who ferment the brew". As a sociologist, Merton dismisses the explanation of scientific disputes being unique expressions of a scientist's nature. He argues that egotism is not a trait of scientists alone, and that the history of social thought has shown this to be a general condition. He also denies the notion that egotistical people are attracted to a scientific career, hungry for fame in a profession that promises such an allure, while acknowledging that there must be some aggressive men of science. Merton makes the point that often it is not the principals, the discoverers of a new truth, who fervently promote such claims, but well-meaning, possibly better-placed friends and champions. For example, Darwin's theory was introduced to the Linnaean Society by Sir Charles Lyell and Joseph Hooker at the same time as Wallace's natural selection theory paper, on 1 July 1858 (Merton, 1957, p. 648; McCalman, 2009, pp. 317–338).

In developing the argument that it is the institution that determines priorities and accolades, Merton writes:

> Recognition for originality becomes socially validated testimony that one has successfully lived up to the most exacting requirements of one's role as a scientist. The self-image of the individual scientist will also depend greatly on the appraisals by his scientific peers of the extent to which he has lived up to this exacting and critically important aspect of his role … recognition and fame then become symbol and reward for having done one's job well (Merton, 1957, p. 640).

He further argues that the most significant form of recognition resides in the practice of affixing the scientist's name to what he has discovered. The naming of Halley's Comet is an obvious example of such acknowledgement. Further down the scale is the recognition of a scientist as the 'father' of a new science or practice. It is noted that France has issued several stamps celebrating their scientists who are recognised as the patriarch of a particular profession and are described as such in popular literature. Figure 5.1 shows three scientists from the 1700s. Lavoisier, the father of modern chemistry, is honoured on the bicentenary of his birth, there is no elaboration of this status, although he is named. Are we to suppose that Lavoisier, during the period of WWII which saw France under German rule, was known well enough not to require more explanation?

I believe we are. The second stamp is from a set of six stamps featuring celebrities (on charity stamps with a premium on the price as a donation to a charity) who are shown with contextual images on either side of the portrait. It has to be said that the indicators of Bichat's pre-eminence in anatomical discovery—he was known as "the father of modern histology and pathology"—are hard to decipher 50 years later. The third stamp shows Fauchard, "the father of modern dentistry", with his life dates given so that we know it is issued on the occasion of the 200th anniversary of his death. His profession is shown as he holds in his hand a thesis on the subject of dentistry.

I am inclined to categorise the two outside stamps in Figure 5.1 as mirrors reflecting a past achievement by a Frenchman. They were both issued as single stamps to mark birth bicentenaries, dates that have come up on the historical calendar. The Bichart stamp is different. Bichart is celebrated in a set of six celebrities within a Red Cross Fund series collecting money for charity through the 10 franc premium being charged on top of the postal rate. The other (all male) celebrants in the set are from mixed professions. As well as Bichart, there is a chronicler, a royal gardener, a philosopher and two sculptors. It is the only time that five of the men have been honoured on French stamps, although sculptor Bartholdi's work has been shown on two art issues. I perceive the Bichart stamp, as the only scientist chosen within this issue and due to the complexity of the symbols—burning and extinguished torches—to be a lens, suggesting there is more of interest to be pursued than would be suggested by a simple portrait of the man.

Figure 5.1: France, 1943, 1959 and 1961. *A-L de Lavoisier (1743–1794), birth bicentenary; Marie François Xavier Bichat (1771–1802), anatomist and physiologist; Pierre Fauchard (1678–1761), death bicentenary.* **Gibbons catalogue # 785, 1432 and 1538.**
Source: Author's collection.

Recognition has also been given to scientists in the past 100 years by the award of Nobel Prizes and these have been a fruitful source of subjects and images for postal authorities. Examples are shown in Figures 5.2 and 5.3. It is interesting to note that the design of the Figure 5.2 stamp is quite sophisticated as it shows the celebrant, the equipment through which his achievement was made, and the iconic image of the Nobel medal. The stamp is another French charity stamp. The purchaser pays a premium 10c on top of the 50c service fee, which is collected by the post office on behalf of the Red Cross Fund. In comparison, the Russian stamps show only the scientist, his name and a very small symbol representing the theme of the issue. Using the same argument as above for Bichart, I see the Grignard stamp as a lens, conveying a combination of symbols illustrating his science successes.

Figure 5.2: France, 1971. *François Auguste Victor Grignard (1871–1935), Nobel Prize-winning chemist* of 1912. Gibbons catalogue # 1915.

Source: Author's collection.

The set of two stamps shown in Figure 5.3 is a mirror. It does not challenge the viewer. It records two Nobel Prize winners during a year that marks Kapitsa's birth centenary and Cherenkov's 90th birth anniversary.

Figure 5.3: Russia, 1994. *Physics Nobel Prize winners*, Pyotr Leonidovich Kapitsa (1894–1984), 1978 Nobel laureate and Pavel Alekseyevich Cherenkov (1904–1990), who shared the Nobel Prize in physics in 1958. Gibbons catalogue # 6491 and 6492.

Source: Author's collection.

Merton draws attention to the perception that Russia was making extravagant claims for its past and current successes in the field of science during the Cold War:

> The recent propensity of the Russians to claim responsibility in all manner of inventions and scientific discoveries thus energetically replicates the earlier, and now less forceful though far from vanished, propensity of other nations to claim like priorities (Merton, 1957, p. 642).

The year 1957 was, of course, at the height of the Cold War, during which the USSR was politically highlighting success in all areas, including science, sport and the arts.

The postage stamp as a device through which to claim a scientific first

As examples of claims to be first, I now consider the discovery or invention of four objects of everyday life and review how history has related the scientific achievements that have led to products we are so familiar with today, and how the postal authorities have told the story.

The light bulb

Thomas Edison's invention of the light bulb was featured on the first United States stamp to celebrate a scientific, technical achievement in 1929 (Figure 5.4). The design of the image looks almost as if a portrait of a personage has been removed and a light bulb inserted in its place. It is the invention that is being celebrated here. The person is almost a given. The main symbol is a light bulb under an arch that declares "United States Postage", with the 2 cent fee replicated in the bottom corners. In the top corners are the dates being celebrated, 1879–1929, described in text as "Edison's Electric Light's Golden Jubilee". The context is clear, although not easy to assimilate. It is a lens, as it requires some thought to decipher, a criterion I apply to a lens.

Figure 5.4: United States, 1929. *The 50th anniversary of the first incandescent lamp*. Scott catalogue # 654.

Source: Author's collection.

In their book *Edison's Electric Light: Biography of an invention*, historians Friedel, Israel and Finn list 22 pioneers of incandescent lamp research prior to Joseph Swan and Thomas Edison. Edison's version is the one remembered because it was part of a fully-working system and was not presented piece-meal (Friedel, Israel and Finn, 1986).

In the early 1800s, Sir Humphry Davy created the first incandescent light by passing current through a thin strip of platinum, chosen because the metal had an extremely high melting point. It was not bright enough nor did it last long enough to be practical.

In 1835, James Lindsay demonstrated a constant electric light at a public meeting in Dundee, Scotland, but having perfected the device to his own satisfaction, he did not develop the electric light any further. None the less, he is credited by Challoner with being the inventor of the incandescent light bulb (Challoner, 2009).

Russian Alexander Lodygin (1847–1923), an electrical engineer and inventor, obtained a Russian patent in 1874 for an incandescent light bulb. As a burner, he used two carbon rods of diminished section in a hermetically sealed glass receiver filled with nitrogen, electrically arranged so that the current could be passed to the second carbon when the first had been consumed. His achievement was celebrated with a stamp in 1951 (Figure 5.5). The wording on the stamp says that he was a renowned Russian scientist and the inventor of the first incandescent light bulb in the world (Windle, 2012). This stamps features in a set of 16 *Russian scientists*, each of which features a mono-coloured, poster-style image. Each of the 16 stamps is a different colour, five of which, including the Lodygin portrait, also show a representation of the scientists' achievement. These stamps are saying that these Russian scientists deserve recognition. The invention is secondary to the person. As a result of the text, this representation is in a lens category.

Figure 5.5: Russia, 1951. *Alexander N Lodygin, inventor of the incandescent light bulb.* **Gibbons catalogue # 1716.**

Source: Author's collection.

From 1850, Joseph Swann (1828–1914), a British physicist and chemist, had been working with carbonised paper filaments in an evacuated glass bulb. By 1860, he was able to demonstrate a working device, but the lack of a good vacuum and an inadequate supply of electricity resulted in a short lifetime for the bulb and it was inefficient source of light. By the mid-1870s, better pumps became available. Swann returned to his experiments and received British Patent No. 8 in 1880.

Thomas Edison's invention of a fully integrated lighting system was demonstrated in 1879 (Hughes, 1977). Edison has been honoured with the release of two other stamps by the United States (Figure 5.6) in addition to the 1929 issue, and by the Republic of Ireland (Figure 5.7). The second US stamp shown does not logically fit into the invention of the light bulb story, but as Edison was such a prolific inventor, it does celebrate his inventiveness. The 1947 portrait that celebrates his birth centenary is again a single colour printing and is one of an issue of five stamps showing *American inventors*, although this particular image commemorates his *Invention of the first phonograph*. Neither of the images in the figure is controversial and I judge them to be mirrors of fact, a recognition of the man and an anniversary in the first stamp and an invention and anniversary in the second.

Figure 5.6: United States, 1947 and 1977. *Thomas Alva Edison, birth centenary and Centenary of the invention of the first phonograph.* **Scott catalogue # 945 and 1705.**
Source: Author's collection.

The Republic of Ireland issue of *New discoveries*, celebrating the millennium, uses images of six featured discoveries (Figure 5.7). The image showing Edison's laboratory and a successful experiment is a larger part of the image, with the name of the set and a millennium symbol shown on the right of the stamp. I have included this image as it purports to show the actual moment when Edison demonstrated incandescence. It is interesting that Ireland has chosen to send such a simple message, the straight-forward celebration of an achievement, to acknowledge the value of the invention. It is a mirror of an event showing the event in context, as more stamp images have done in the year 2000 as world science is viewed from a historical perspective. There is no overt claim of being first, it is an acknowledgement of achievement that has made living easier for everybody.

Figure 5.7: Éire, 2000. *Millennium Issue III – New Discoveries*, *Thomas Alva Edison*, *the incandescent lamp*, **from a set of six. Hibernian catalogue # C982.**
Source: Author's collection.

After the reunification of East and West Germany, the postal authority issued a stamp celebrating the *150th anniversary of the light bulb* and naming Heinrich Göbel (1818–1893), as the inventor. Subsequent research has shown that this German-born physicist moved to the United States and did work upon the development of the light bulb powered by a battery produced from zinc and carbon elements. In 1854, his experiment became a reality, and the electric bulb gave light to mankind for the first time in history. When Edison took legal action to protect his patents, a defence of Göbel's case was mounted and was successful (Tanner, 1894).

The German stamp issue shown in Figure 5.8 celebrates Göbel, naming the anniversary and the picture of what it calls the "Göbelampe 1854", as well as illustrating its modern equivalent. This is another stamp, similar to the first Edison stamp in Figure 5.1, in which the inventor is subsidiary to the invention, although Göbel's name is appended to the name of the lamp. The Edison US claim is for an 1879 invention. Germany says Göbel did it first and showed it in 1854. The message challenges. It is a lens, partly, I believe, because of the image representation of the evolution of the technology.

Figure 5.8: Germany, 2004. *Heinrich Göbel, 150th anniversary of the light bulb*. Gibbons catalogue # 3266.

Source: Author's collection.

The telephone

Several claimants have been promoted on stamps seeking the substantiation of who invented the telephone. The twist in the tail is that in 2001 the United States Congress passed a motion stating that it had actually been an Italian, Antonio Meucci, who should be credited with that accolade. Prior to this motion, the main argument had been about whether Alexander Graham Bell or Elisha Grey was first. In the story of the telephone, the February 1876 machinations of these two scientists with the US Patent Office show how close the race was (Catania, 2002). Catania states that the US Government claimed Antonio Meucci and Philipp Reis to be the progenitors of all subsequent telephone models. In particular, the US Government asserted that the make-and-break transmitter devised by Philipp Reis could be adjusted to work as a microphone, namely as a "variable resistance transmitter". The controversy surrounding the invention of the telephone has allowed nations to state their case, with or without evidence.

It was Italian Innocenzo Manzetti (1826–1877) who first mooted the idea of a "speaking telegraph", and he built his first device in 1864. This work was acknowledged by fellow Italian, Antonio Meucci. During 1861, Johann Philipp Reis (1834–1874) publicly demonstrated the Reis telephone before the Physical Society of Frankfurt. Reis's work was complementary to the research being conducted by Meucci. At the time, Reis was devising receivers and make-break transmitters using magnetostriction (Catania, 2002). Some celebratory stamps are shown in Figure 5.9. This is the first time I have shown a time sequence of stamps from the same country celebrating the same person and his achievements. The image trend is from portrait, to an image of the equipment, and the third stamp has a likeness of Reiss using a telephone in front of a simple telephone switchboard. The trend is towards showing more context, explaining the relationship of the scientist to his invention. In this case, the three images are a move from mirrors to lens as the story develops.

Figure 5.9: West Germany 1952, 1961 and 1977. Philipp Reis on the anniversary of *75 years of the German Telephone Service*; the *Centenary of the first demonstration*; and *100 Years of Service*. Gibbons catalogue # 1087, 1287 and 1837.

Source: Author's collection.

Figure 5.10, a later West Germany issue, shows Reis and his device, the first make-and-break system that he demonstrated to the Royal Prussian Telegraph Corps in 1862. The emphasis is on the man, although the equipment is shown. The German text names Reis and states "erfindung des telefons": inventor of the telephone. The written claim together with the scientific context classifies this image as a lens.

Figure 5.10: West Germany, 1984. *150th birth anniversary of Philipp Reis.* Gibbons catalogue # 2048.

Source: Author's collection.

East Germany, although politically opposed to West Germany at the time, has twice honoured Philipp Reis (Figure 5.11). The first image is a stylised representation of a 1970s home telephone. The second is from a set of four, titled *Telephones,* with the Reiss phone being the first in chronological order. The images seem particularly detailed and clear mirrors of reality.

Figure 5.11: East Germany, 1976 and 1989. *Centenary of the German Telephone System* **and** *The 1861 Reiss telephone*. **Gibbons catalogue # E1833 and E2927.**

Source: Author's collection.

During 1865, La Feuille d'Aoste reported: "It is rumored that English technicians to whom Mr Manzetti illustrated his method for transmitting spoken words on the telegraph wire intend to apply said invention in England on several private telegraph lines."

In 1871, Antonio Meucci filed a patent caveat, titled "Sound Telegraph", in the US Patent Office describing communication of voice between two people by wire. His first equipment for voice transmission was built in 1857 and finalised in 1870. This electromagnetic product performed clear transmission of speech and included fundamental techniques for long distance communication (Catania, 1994). However, after having renewed the caveat for two years, Meucci failed to find the money to renew it. The caveat lapsed and Alexander Graham Bell's US Patent 161,739, "Transmitters and Receivers for Electric Telegraphs", was granted in 1875. This used multiple vibrating steel reeds in make-break circuits (Bruce, 1990). Bell's contribution is commemorated on the stamps shown in Figure 5.12. The first image is from a set of five *American inventors* issued in 1940, at which time the United States was feeling its way before it entered WWII. In the same year, the US Postal Service also issued sets of five authors, poets, educators, scientists, composers and artists. Bell is shown in a simple portrait which takes most of the available space, with his name, value and the crest of the American Academy of Arts and Science in an inset box. The image on the second stamp, issued on the *Centenary of the US telephone system*, is a reproduction of the drawing from Bell's 1876 patent. His name and the anniversary details are also shown. The trend is two-fold, as the images move from mirror to lens and from simple portrait to a quite detailed drawing of how a telephone works that attributes the invention to Bell. In using the technical drawing, a strong claim is being made that it was Bell who invented the telephone in 1876.

Figure 5.12: United States, 1940 and 1976. *Alexander Graham Bell* and *Centenary of the US telephone system.* Scott catalogue # 893 and 1683.
Source: Author's collection.

On 14 February 1876, Elisha Gray filed a patent caveat for transmitting the human voice through a telegraphic circuit. The same day, Alexander Bell applied for the patent "Improvements in Telegraphy", for electromagnetic telephones using undulating currents. Five days later, Gray was notified by the US Patent Office of interference between his caveat and Bell's patent application. Gray decided to abandon his caveat. On 7 March, Bell's US patent 174,465, "Improvement in Telegraphy", was granted, covering "the method of, and apparatus for, transmitting vocal or other sounds telegraphically … by causing electrical undulations, similar in form to the vibrations of the air accompanying the said vocal or other sound" (Brooks, 1976). The first successful telephone transmission of clear speech using a liquid transmitter occurred when Bell spoke the words, "Mr. Watson, come here, I want to see you", into his device and Watson heard each word distinctly (Bruce, 1990).

On 30 January 1877, Bell's US patent was granted for an electromagnetic telephone using permanent magnets, iron diaphragms, and a call bell. Thomas Edison, the serial inventor, filed for a patent on a carbon (graphite) transmitter. His patent was granted in 1892, after a 15-year delay because of litigation. Edison was also granted a patent for a carbon granules transmitter in 1879. The literature relates the ongoing assessment between the claims of Gray and Bell until the twenty-first century, but both are now superseded by Meucci, in the US at least.

During 2002, the American Congress recognised Meucci, not Bell, as the inventor of the telephone: "Resolved, that it is the sense of the House of Representatives that the life and achievements of Antonio Meucci should be recognized, and his work in the invention of the telephone should be acknowledged" (US House of Representatives Resolution 269)

Neither Innocenzo nor Antonio Meucci have been celebrated on the stamps of Italy.

The wireless

The work of the scientists contributing to radio's prehistory during the nineteenth century has also been recognised. A chronology has been derived from the same sources as were used to look at the development of the telephone.

In 1820, Hans Christian Ørsted discovered the relationship between electricity and magnetism in a very simple experiment. He demonstrated that a wire carrying a current was able to deflect a magnetised compass needle. In Great Britain during the 1830s, Michael Faraday began a series of experiments in which he discovered electromagnetic induction: that a changing magnetic field could produce a current. The relation was mathematically modeled by Faraday's law, which subsequently became one of the four Maxwell equations. Faraday proposed that electromagnetic force fields extended into the empty space around the conductor, but did not complete his work involving that proposal. Faraday has been celebrated twice on the stamps of Great Britain. Figure 5.13 shows a profile portrait of him thinking creatively with a diagram of an apparatus as a celebration of the bicentenary of Faraday's birth. The indexation of the image shows a profile of Queen Elizabeth II, the value 22p, and Faraday's name linked to an association with electricity. Reading the messages conveyed with this image is not easy. Another 22p stamp in the set shows Charles Babbage thinking numbers in his head, whilst the two other stamps in the set show diagrams of radar, celebrating the 50th anniversary of Sir Robert Watson-Watt's invention of radar, and a jet aircraft commemorating the 50th anniversary of Sir Frank Whittle's jet engine. Deciphering the messages makes the set a lens, requiring some mental effort, although the celebrations are simple to understand.

Figure 5.13: Great Britain, 1991. *Birth centenary of Michael Faraday*, one of a set of four entitled *Scientific achievements*. Gibbons catalogue # 1546.

Source: Author's collection.

Between 1861 and 1865, James Clerk Maxwell's work was essentially theoretical. His key early paper of the time was "On physical lines of force". Later, in 1873, as a result of experiments, Maxwell first described the theoretical basis of the propagation of electromagnetic waves in his paper to the Royal Society, "A Dynamical Theory of the Electromagnetic Field".

In the United States in 1875, physical experimenter Thomas Edison announced to the press that while experimenting with the telegraph he had noted a phenomenon that he termed "etheric force". He abandoned this research when Elihu Thomson, among others, ridiculed the idea.

In 1878, David E. Hughes was the first to transmit and receive radio waves when he noticed that his induction balance caused noise in the receiver of his homemade telephone. David Hughes demonstrated his discovery to the Royal Society two years later, but was told it was merely induction. Just a few years later, in 1884, Temistocle Calzecchi-Onesti at Fermo in Italy invented a tube filled with iron filings, called a "coherer". From 1884 to 1886, Edouard Branly of France produced an improved version of the coherer. Branly has been commemorated on two stamps (Figure 5.14). Printed in 1944, the first, printed in blue, names Branly and shows his life dates as 1844–1940, and gives the reason for his commemoration being his *Birth centenary*. The background to the portrait is shaded, but the name of the designer, Decaris, is just visible. The stamps of France at this time sometimes included the designer's name. The second stamp, printed in chocolate colour is one of a set of six printed and sold at a charity premium in 1970, in the series named the *Red Cross Fund*, which mainly featured French celebrities. The stamp shows the subject's name in full and repeats the life dates. Branly's portrait has context added as he looks into a village scene surrounded by the outline of an electrical circuit and three items of equipment dependent upon the circuit. The name of the designer, Serveau, takes a magnifying glass to decipher. Branly is shown in the context of his fame. The blue stamp is a mirror, the other a lens.

Figure 5.14: France, 1944. *Birth centenary of Edouard Branly. 1970,* from a *Red Cross (Charity)* set of celebrated persons. Gibbons catalogue # 811 and 1860.

Source: Author's collection.

In 1885, Edison took out a patent on a system of radio communication between ships, which he then sold to Guglielmo Marconi, who is featured on stamps shown in Figure 5.20. Between 1886 and 1888, Heinrich Rudolf Hertz (1857–1894) validated Maxwell's theory through experiment. He demonstrated that radio radiation had all the properties of waves (now called Hertzian waves). The Hertz experiments are regarded by physicists as definitive regarding electromagnetic waves. Hertz has been featured on a number of stamps, three of which are shown in Figure 5.15. The third stamp, a German stamp of 1994, commemorates Hertz's *Death centenary*. His portrait is shown against a background of electromagnetic wave forms. Again requiring a microscope to read, the year of issue, 1994, is shown in very small figures beneath the portrait. Showing the issue date has been a later requirement of the Universal Postal Union. The first two birth centenary stamps show not much more than a portrait, although both show his name and life dates. They celebrate the man rather than his achievements. The middle stamp is from an East German set of three, entitled *Scientists anniversaries,* and is in the same style as the 1950 definitive set that celebrated the *250th anniversary of the Academy of Sciencesin Berlin*. From left to right, we have two mirror and one lens image. The trend to provide context is again apparent with time.

Figure 5.15: West Germany, 1957. *Birth centenary of Heinrich Hertz*: **DDR, 1957.** *Birth centenary of Heinrich Hertz*; **Germany, 1994.** *Death centenary of Heinrich Hertz*. **Gibbons catalogue # 1178, E324 and 2557.**
Source: Author's collection.

There are multiple claims to the invention of radio. Key developers included Marconi, who equipped ships with life-saving wireless communications and established the first transatlantic radio service. Tesla developed the means to reliably produce radio frequencies, publicly demonstrated the principles of radio, and transmitted long distance signals. At St. Louis, Missouri in 1893, Tesla gave a public demonstration of wireless radio communication. Addressing the Franklin Institute in Philadelphia and the National Electric Light Association, he described in detail the principles of radio communication. The apparatus that he used contained all the elements that were incorporated into radio systems before the development of the "oscillation valve", the early vacuum tube. Tesla was the first to apply the mechanism of electrical conduction to wireless practices. He initially used sensitive electromagnetic receivers that were unlike the less responsive coherers later used by Marconi and other early experimenters. Afterwards, the principle of radio communication was publicised widely and various scientists, inventors, and experimenters began to investigate wireless methods. Tesla was celebrated by the United States as an *American inventor* in 1983 (Figure 5.16). The stamp names him along with a drawing of an induction motor and his likeness.

Figure 5.16: United States, 1983. *Nicolas Tesla*, **from set of four,** *American Inventors*. **Scott catalogue # 2057.**
Source: Author's collection.

British physicist Sir Oliver Lodge demonstrated the reception of Morse code signaling using radio waves using a "coherer" in 1894. That same year, Indian physicist Jagdish Chandra Bose demonstrated publicly the use of radio waves in Calcutta, but he was not interested in patenting his work. Bose ignited gunpowder and rang a bell at a distance using electromagnetic waves, proving that communication signals can be sent without using wires. Two years later, in 1896, Bose went to London on a lecture tour and met the Italian Marconi, who was conducting wireless experiments for the British Post Office.

Almost concurrently, in 1894 the Russian physicist Alexandr Stepanovich Popov (1859–1906) built a coherer, his first radio receiver. He has received a continuing recognition for his work from the Russian stamp authorities. He was the first to demonstrate the practical application of electromagnetic (radio) waves, although he did not apply for a patent for his invention. Beginning in the early 1890s, he continued the experiments of other radio pioneers, such as Heinrich Hertz. Further refined as a lightning detector, his receiver was presented to the Russian Physical and Chemical Society on 7 May 1895. The day has been celebrated in Russia since then as Radio Day. In March 1896, he effected transmission of radio waves between different buildings in St Petersburg. Upon learning about Guglielmo Marconi's system, he was able to effect ship-to-shore communication over a distance of 6 miles in 1898 and 30 miles in 1899.

Popov is described in historical texts as a Russian hero of science. Popov has been celebrated and honoured on Russian stamps almost every decade since the 1920s. His birth and death anniversaries (occuring in 1925, 1949, 1955, 1959, 1965, 1972, 1989, 1995 and 2009) are reason enough for another stamp issue. Russia shows a real loyalty to its heroes once they are recognised. Shown below are three of the nine issues listed as Figures 5.17, 5.18 and 5.9. They show how the portrayal of a scientific celebrity has changed over time. Figure 5.17 shows a set of three, single colour portraits of Popov, two with him standing before a blackboard clearly illustrating an electrical circuit, and the other as a full face image. All show text the reason for the issue: the *50th anniversary of Popov's radio discoveries*. The man is honoured more than the invention with a full-face representation, seemingly the only image of him. These three stamps are mirrors showing the man in factual terms.

Figure 5.17: Russia, 1945. *50th anniversary of Popov's radio discoveries*. **Gibbons Catalogue # 1114-1116.**

Source: Author's collection.

Issued 20 years later, Figure 5.18 is a miniature sheet that shows a selection of six applications dependent upon Popov's radio discoveries. Somewhat unusually, the individual stamps in the sheet do not have a face value. It is presumed that the Russian Postal Authority anticipated that the sheet, face value 1 ruble, would be used fiscally in its entirety. The images are Popov's initial radio invention, a transistor radio, a television screen, radar, a radio telescope and a telecommunications satellite. The six stamps are tied together from a design perspective with a small symbol showing the 70 year celebration. As examples of Popov's achievements in six different applications, the miniature sheet is a lens. I perceive the sheet has been printed more as a label as the individual stamps do not carry a value, so the expectation is that the sheet remains intact. The high value suggests it has been fabricated to appeal to specialist collectors. Paradoxically, the text is Russian and you would have to know about Popov to appreciate the purposes to which his discoveries, as illustrated, have been utilised in 1965. This is a mixed message, therefore, and a lens.

Figure 5.18: Russia, 1965. *70th anniversary of Popov's radio discoveries.* **Gibbons catalogue # MS3135.**

Source: Author's collection.

Figure 5.19 was issued in 2009 to mark Popov's 150th birth anniversary. The stamp is one of very few Russian stamps that have been perforated to allow for a round stamp. It is a high value stamp, at 20 rubles face value. In its form as a miniature sheet, it conveys a lot of content. The title of the sheet is shown as though printed by a teleprinter. The background envisages radio waves emanating from Popov and includes detailed images of a circuit diagram, the original radio equipment, and notes in Popov's hand. The stamp, were it to be removed from the sheet, shows (again the only) portrait used by Russia for Popov, along with his details, the country name in Cyrillic and English, and the year of issue. The presentation of person and his invention with such clarity marks it as a lens issue.

5. On Being First

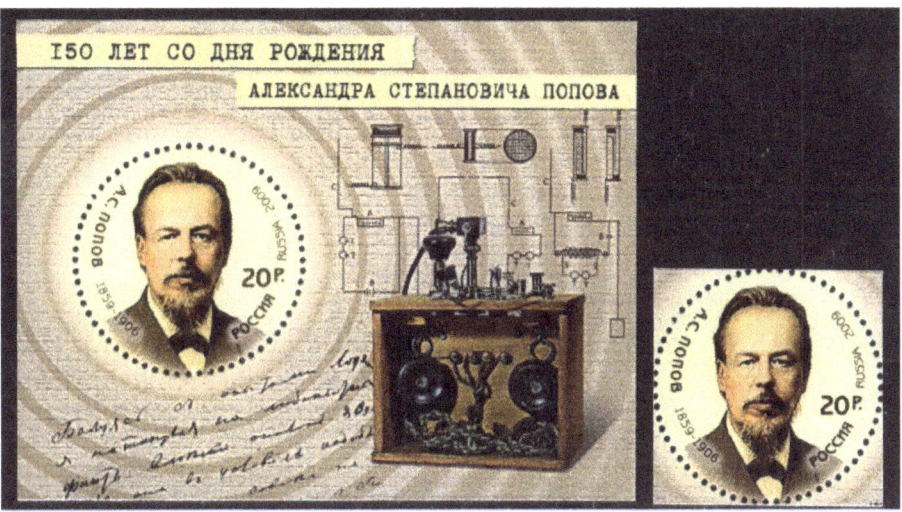

Figure 5.19: Russia, 2009. 150th birth anniversary of the inventor of radio, *Aleksandr S. Popov*. WNS numbering system # RU011.09.
Source: Author's collection.

In 1896, Gugliemo Marconi—who is shown in two Great Britain stamps in Figure 5.20—was awarded a patent for radio with British Patent 12039, "Improvements in Transmitting Electrical Impulses and Signals, and in Apparatus therefor". This is the initial patent for radio, though it used earlier techniques of various other experimenters, primarily Tesla, and resembled the instrument demonstrated by others, including Popov. During this time, spark-gap wireless telegraphy was widely researched. The first stamp, the 41p, shows a young Marconi and celebrates "Marconi first wireless message 1895", the words shown on the bottom of the stamp. The background image is a map of the North Atlantic, which was Marconi's testing ground, while the equipment used appears in the foreground. What is interesting is that the same photograph has been used for the later stamps (see Figures 5.23 and 5.24). The second stamp, 60p value, shows an older Marconi using a telephone handset, and has the same descriptive message as the 41p stamp. The background image now carries an S.O.S. call over the image of a sinking ship, presumably the Titanic. Based upon photographs, the images are active portrayals of Marconi rather than portraits. Both stamps are contextually rich and are lenses into the invention.

Figure 5.20: Great Britain, 1995. *Guglielmo Marconi, pioneer of communications*, **from a set of four. Gibbons catalogue # 1889 and 1890.**
Source: Author's collection.

During 1897, Marconi established a radio station on the Isle of Wight, England. In the same year, in the US, Tesla applied for two key radio patents, which were issued in early 1900. The text on the stamp in Figure 5.21 states it that it shows the "Marconi/Kemp experiments 1897". An oscillator and spark transmitter are shown. This stamp comes from a set of four, entitled *Broadcasting anniversaries*. The stamp is a lens, as the method of use of the equipment is not illustrated, and also for the fact that Kemp, Marconi's assistant in the early 1900s, gets a mention.

Figure 5.21: Great Britain, 1972. *Marconi/Kemp experiments 1897.* **Gibbons catalogue # 912.**
Source: Author's collection.

Marconi claims to have received a radio signal transmitted from Poldhu in Cornwall (UK) in St. John's, Newfoundland, in 1901, although this is disputed. The claims of Marconi's signal and case against it are still discussed. In 1904, the US Patent Office reversed its decision, awarding Marconi a patent for the invention of radio, possibly influenced by Marconi's financial backers in the US, who included Thomas Edison and Andrew Carnegie. This also allowed the US Government, among others, to avoid having to pay the royalties that were being claimed by Tesla for use of his patents.

Marconi's achievements have been recorded by four of the ten countries in this study. The United States identified Marconi's spark coil and spark in its 1973 *Progress in Electronics* issue (Figure 5.22). The message of the three stamps, all with illustrations of electronic equipment, was to record development towards transistors and a printed circuit board, which was the current technology of the 1970s.

Figure 5.22: United States, 1973. *Progress in electronics*. Scott catalogue # 1500.

Source: Author's collection.

Marconi has also been shown on stamps by Reunified Germany in 1995, and by Ireland in 1995, celebrating the centenary of the first radio transmission and recognising him as the inventor of radio (Figures 5.23 and 5.24). Both stamps use a classic photograph of Marconi with his equipment. The stamps are so similar that one suspects the two postal authorities have colluded in the design. Great Britain used the same image in Figure 5.20. The Irish stamp was issued se-tenant, as shown, with a stamp of the same value showing a radio dial. Both stamps are mirrors, a note on history, as the images are familiar. However, the Irish stamp may be the more pertinent, as Marconi's major transmission achievements were conducted from West Ireland, from the major purpose-built first transmitter, so perhaps also a lens perspective, although I have not been able to find this fact recorded philatelically.

Figure 5.23: Germany, 1995. *Guglielmo Marconi, centenary of first radio transmission*. Gibbons Catalogue # 2638.

Source: Author's collection.

Figure 5.24: Éire, 1995. *Centenary of radio, Guglielmo Marconi*. Hibernian catalogue # C717 and C718.
Source: Author's collection.

A 1974 West Berlin stamp has introduced an interesting character (Figure 5.25). As a result of his personal acquaintance with Sir William Henry Preece, the head of the English telegraph administration, German engineer Adolf Slaby participated with the help of his assistant Georg von Arco, in Marconi's experiments with wireless telegraphy across the English Channel from 1897 (Vyvyan, 1974). Vyvyan writes that Slaby recognised immediately the possibilities, and repeated the experiments in Berlin, winning the interest of the government and the military. This stamp is understated, with the named portrait and a background representation of wireless waves. It has to be assumed that the issuing authority expected the public to know the name and achievements of Adolf Slaby. The stamp is a mirror; it does not appear to challenge unless you know some of the story.

Figure 5.25: West Berlin, 1974. *The 125th birth anniversary of Adolf Slaby (1849–1913)*. Gibbons catalogue # B450.
Source: Author's collection.

Flight

The history of flight has been faithfully recorded by postal authorities with the implied, and sometimes again stated, fact that their claimant is deserving of the acknowledgement of having achieved a first in the world history of flight.

Furvin Kryakutnoy was an early-eighteenth century Russian inventor. For a while it was thought he may have invented the hot air balloon 50 years before the Montgolfier brothers. Figure 5.26 shows an image which may honour Kryakutnoy. However, the Gibbons' Russia catalogue notes that "there is considerable doubt over the identity of the balloonist involved ... and the date of the event" (Stanley Gibbons Publications, 2010, part 10, p. 73). As the stamp challenges thought, it is a lens.

Figure 5.26: Russia, 1956. 225th anniversary of first balloon flight in 1731 by Furvin Kryakutnoy. Gibbons catalogue # 2034.
Source: Author's collection.

In 2006, France issued a set of six stamp, *Flying machine*, which included a stamp celebrating the achievements of Jean-Marie Le Bris (1817–1872) (Figure 6.27). Le Bris, a naval officer, built a glider inspired by the shape of the albatross and named *L'Albatros artificiel* ("the artificial albatross"). During 1856, he flew briefly on the beach of Sainte-Anne-la-Palud (Finistère), by being pulled by a running horse, facing towards the wind so that people could not say he flew using the wind. He flew higher than his point of departure, , reportedly to a height of 100 metres for a distance of 200 metres, a first for heavier-than-air flying machines. The illustrations of the glider shapes invoke, in this observer, a fascination for the time and the events: this stamp is a lens.

Figure 5.27: France, 2006. *Flying machines*, from a set of six, *Jean Marie Le Bris*. WNS numbering system # FR120.06.
Source: Author's collection.

The East German Democratic Republic staked its claim in 1991, publishing a set of four stamps with the title *Historic flying machines designs,* including Da Vinci sketches. Otto Lilienthal (1848–1896) was a German pioneer of human aviation who became known as the Glider King. He was the first person to make well-documented, repeated, successful gliding flights. He followed an experimental approach established earlier by Sir George Cayley. Newspapers and magazines published photographs of Lilienthal gliding, favorably influencing public and scientific opinion about the possibility of flying machines becoming practical. The German Glider King was a pioneer of human aviation, the first person to make controlled untethered glides repeatedly and the first to be photographed flying a heavier-than-air machine. He made about 2,000 glides prior to his death on 10 August 1896 from injuries taken during a glider crash the day before. Figure 5.28 shows a Lilienthal glider from two perspectives and names him as the celebrated designer. This is another stamp full of intriguing detail, again a lens.

Figure 5.28: DDR, 1990. *Historic flying machines design*, *Otto Lilienthal*. **Gibbons catalogue E3010.**
Source: Author's collection.

Figure 5.29 shows two stamps from an airmail set of 11 with the portraits and the images of the flying machines of Lilienthal and von Zeppelin, another German celebrated in history for his airships. These two stamps are very early examples of a scientific or technical context being added to the portrait that dominates the image. During my study, the trend emerges whereby the stamp image changes from the person being the pre-eminent focus of the message to the context becoming equally important, adding an awareness factor to the message. These are mirrors, given the dominance of the portrait and the familiarity of the images of glider and airship.

Figure 5.29: Germany, 1934. *Otto Lilienthal (1848–1896)* and *Count Ferdinand von Zeppelin (1838–1917)*. Gibbons Catalogue # 535 and 536.
Source: Author's collection.

During the period 1948–1990, the West Berlin postal authority maintained a careful equilibrium with respect to Cold War politics and issued a set of four stamps in 1978 celebrating *Aviation History*, that included Otto Lilienthal and the Wright Brothers on equal footing. The images show Lilienthal under a set of glider wings, dated 1891, and bi-plane model of the Wright Brothers, dated 1909 (Figure 5.30). All the stamps in the set name the celebrant and subject.

Figure 5.30: Germany, 1978. *Otto Lilienthal* and *the Wright Brothers*. Gibbons catalogue # 8548 and 8549.
Source: Author's collection.

Australia also has a candidate for the first flight by a heavier-than-air vehicle. In 1994, Australia Post issued a set of four stamps celebrating aviation feats. Three of the celebrants were aviators, but Lawrence Hargrave (1850–1915) was an experimental design scientist. The Australian inventor of the box-kite linked four of his kites together, added a sling seat, and flew 16 feet in 1894. By demonstrating to a skeptical public that it was possible to build a safe and stable flying machine, Hargrave opened the door to other inventors and pioneers. The achievement is shown in Figure 5.31. The design is rather busy, incorporating box-kite designs from two perspectives, early wing designs, a portrait, and the naming of Hargrave. The date of the flight is shown in large letters but in a rather neutral colour from the background design. The design is modern, emphasising both the man and his machine. This is a page of history, perhaps, not well known to the Australian public and therefore is a lens.

Figure 5.31: Australia, 1994. *Centenary of an aviation feat*, Lawrence Hargrave. Renniks' catalogue # 1441.

Source: Author's collection.

Hargrave devoted most of his life to constructing a machine that would fly. He believed passionately in open communication within the scientific community and would not patent his inventions. Instead, he scrupulously published the results of his experiments in order that a mutual interchange of ideas may take place with other inventors working in the same field, so as to expedite joint progress. While the Wright brothers denied that they owed anything to Hargrave, his discovery of the cellular kite and his investigations into the superiority of curved wing surfaces played an important part in European experimental work, which culminated in the first public flight by Santos-Dumont in France in 1906.

Alberto Santos Dumont (1873–1932) appears within the same French set of six *Flying machines* as Le Bris. The design incorporates four perspectives of the vehicle he demonstrated, named Demoiselle, and Dumont's name. Figure 5.32 combines the attributes of a mirror and also a lens into the history of flight.

Figure 5.32: France, 2006. *Flying Machines, Alberto Dumont*. WNS numbering system # FR119.06.

Source: Author's collection.

Within the millennium issue of New Zealand is the challenging stamp shown in Figure 5.33. It was issued as a part of a set of six in 1999 and included in a miniature sheet in 2000. The stamp celebrates the inventiveness of Richard William Pearse (1877–1953). The text describing the stamp in the *1999 Yearbook of New Zealand Post*, intrigues; "Though the first recognised flight by a heavier-than-air machine was that of the Wright Brothers in Flyer 1 on 17 December 1903 that flight may not have been the first" (Di Somma, 1999, p. 66). This millennium stamp uses as its main image the drawing of Pearse's airplane. Telling the message are a logo of New Zealand, used for the millennium issues, and the words "leading the way" and "powered flight c. 1903". This stamp challenges the status quo, and is a lens.

Figure 5.33: New Zealand, 1999. *Millennium issue* (Series V), *Leading the way, powered flight c. 1903*. Campbell Paterson catalogue # SH104.
Source: Author's collection.

George Bruce Bolt (1893–1963), an aircraft pioneer and chief engineer of Air New Zealand, undertook initial research into the achievements of aviation pioneer and inventor Richard Pearse during the late 1950s and early 1960s. More recently, Geoffrey Rodliffe has pursued the Pearse story and the claim that his flights pre-dated those of the Wright Brothers (Rodliffe, 2008). Pearse is shown in Figure 5.34 in a stamp from a set of six *New Zealand achievers*. It is an image very much like the Lilienthal image in Figure 6.29. It is the man who is celebrated, his machine is added to provide the context; this is a lens because of the controversial claims made for Pearse:

> Pearse must be the first and the only aviator who had at that time designed his own unique internal combustion engine; and designed his own aircraft (pre-dating the microlight by about seventy years). He built both the engine and aircraft in his secluded elementary workshop with minimal assistance, enabling him to make a number of flights piloting the aircraft himself (Rodliffe, 2008).

Figure 5.34: New Zealand, 1990. *The Achievers, Richard Pearse.* **Campbell Paterson catalogue # SH28.**
Source: Author's collection.

The United States has celebrated the Wright Brothers' initial controlled flight three times since 1928. The first time was to mark the International Civil Aeronautics Conference held in Washington, DC during December 1928. The issue also served as the 25th anniversary of the Wrights' first flight of 17 December, 1903. The format of the stamps follow United States Postal Service convention of the times. Figure 5.35 includes "US Postage" as an index, and the value (in figures in the bottom corners) constitute the indexation. The name of the conference is also given, as are the conference dates. The location shown with Congress and Washington Memorial images either side of images of airplanes. The 2 cent stamp shows Kitty Hawk, the Wright's vehicle, and a later biplane superimposed over a globe to represent the international aspect of the conference. The stamps are mirrors of the event.

Figure 5.35: United States, 1928. International Civil Aeronautics Conference, Washington and 25th anniversary of the Wrights' first flight. Scott catalogue # 649 and 650.
Source: Author's collection.

Sketch representations of Orville and Wilbur Wright are shown on three stamps of the United States in 1949 and 1978. Figure 5.36 was issued as a part of the *Millennium series* of US stamps. The stamp is worth looking at from two perspectives. First, it is an example of very few stamps that have been printed with an explanation of the main image on the obverse side. Second, the way the image has been drawn seems to show a flight that reached several hundred feet. The first flight did not. The stamp is a lens.

Figure 5.36: United States, 1998. *Celebrate the century, the 1900s*. The front and back of the stamp celebrate Kitty Hawk and the Wright Brothers. Scott catalogue # 3182g.
Source: Author's collection.

Figure 5.37 seems to take the Wright Brothers first flight even higher, to a height where the earth's curvature is clearly visible. This stamp celebrates the centenary of the first flight. It is a mirror of the event.

Figure 5.37: United States, 2003. *Centenary of first flight*. Scott catalogue # 3783.
Source: Author's collection.

It is clear that in pursuing their objectives of telling their nations' stories, particularly historical stories, postal authorities look to scientific achievement as a viable conduit for their messages. Aircraft provide such a vehicle in an immediately recognisable context. None of the claims examined has been beyond reason. Such claims are made and substantiated through publication and remain in circulation for as long as stamps are saved in collections and are included in the annual update of the country's catalogue of stamp issues.

To conclude this section of my analysis, I need to add a post-script. Merton (1957) describes the most extreme level of deviant behaviour, that of seeking to be recognised as being the first in science, through fraud. To illustrate one example of this behaviour my study needs to step outside of the ten countries being examined, to South Korea (Figure 5.38). I have to acknowledge that the dubious nature of the celebration was unknown at the time of issue. The stamp shown in Figure 5.38 is semiotically rich. The design incorporates as indices the South Korea name and emblem, the year of issue and achievement as 2005. The left-hand side of the stamp incorporates laboratory equipment, the title of the issue, and a magnified illustration of a medical procedure involving a stem cell. But what makes this a hard-edged lens is the sequence of events, which shows a disabled person in a wheelchair, presumably after treatment with human-cloned embryonic stem-cells, through four phases of recovery to a place where a normal life is possible.

Figure 5.38: South Korea, 2005. The successful establishment of human cloned embryonic stem cells. WNS system # KR005.5.
Source: Author's collection.

In its January 2006 issue, in a section called "science in culture", *Nature* magazine looked at the story behind the falsification of stem-cell research of Korean scientist Woo Suk Hwang, and the issue of the stamp shown above, in an article entitled "Stamping his Authority: The Hwang scandal highlights the dangers of hyping science". Martin Kemp, the author of the article, concludes by stating: "the question raised by the stamp and other such visual and verbal hype is whether it is now possible to become a big beast in the jungle of science without becoming ensnared in the perilous mechanisms of celebrity" (Kemp, 2006, p. 396). Kemp endorses Merton's assertion about institutions promulgating priorities when he writes:

> Hwang's rise involved celebrity-minded scientists, state bodies in South Korea concerned with national prestige, funding agencies accountable to government masters, educational institutions bent on international competition, and journalists intent on good stories. They came together in a complex symbiosis to create a distorted image of scientific achievement (Kemp, 2006, p. 396).

Merton's 1957 paper has been shown to be relevant today and, indeed, the images and stories described in this section could well be used to illustrate his paper. Looking at the question of who was first and the scientists featured on postage stamps through the mirror or lens perspective shows that the scientist and his achievement can be used either as a reflection of reality or as a prompt to further enquiry or thought. It depends on the overall message the designer wants to tell.

A University of Pittsburgh paper prompts me to add an additional thought concerning the visual representation of science and the use of images:

> Given the prevalence of visual representations in scientific discourse, it begins to seem that if these visual representations are to be taken seriously as genuine parts of scientific discourse, they will have to be understood to express claims, (Visual representations in science (abstract) philsci-archive, 2006).

Given that this study is seeking to understand the representation of science and scientists on postage stamps, I endorse the proposition that they are genuine parts of scientific discourse. In the next chapter, I examine the heroes of science celebrated on stamps.

6. Scientists on Stamps

Heroes of science

Two quotations outline my thoughts behind looking at heroes of science:

> … stamps constitute the tip of the iceberg of the nexus of cultural, historical and political forces of the society to which they give expression … offering images which provide the possibility of a degree of independent or national assertion (Scott, 1995, p. 94).

> Postage stamps are particularly suitable for making comparisons between attitudes, (to science and scientists), in different countries, because they communicate chiefly in pictorial or symbolic fashion, no great linguistic skill is needed to interpret their messages (Jones, 2001, p. 404).

The analysis of Chapter Three clearly supports the ideas put forward in these quotations. This chapter now seeks to explore in further detail the way that they key scientists are portrayed on stamps. The title "heroes of science" implies that the scientists concerned have had their reputations enhanced, that in some way they have become celebrities, this acknowledged through their exposure on stamps. In Chapter Three, I described the reasoning behind my inclusion of explorers as scientists in this study. Newer countries that have themselves been explored record the experience of exploration as part of their immediate history, as well as celebrating subsequent anniversaries of such events. By implication, therefore, foreign scientists have contributed to the history of the country being explored. This has been a common thread throughout my study. It is also clear that history, as recorded on postage stamps, may continue to cause distress to subjugated indigenous populations, but this is largely ignored. Some reparation has been made through acknowledgement of these different cultures on modern postage stamps but not (yet) through science.

In the review that follows, I shall mention well known explorers, but largely concentrate upon examples of local-born heroes. Examples have been chosen because they feature prominently in the country of choice over a long time period, which allows for additional comparison of design and context.

The taxonomy of the stamps in this study shows approximately half of all science stamps using the image of a named scientist as the main feature. I have recorded stamp images of those scientists who have been celebrated on the country's stamps more than once in country and chronological order. I have also paid attention to those foreign scientists honoured by a country that does not claim them as their own. What are the reasons for portraying scientists on stamps? An association with an achievement is obvious. It might also be used as a nation-building or a civic education objective, or as a feel-good initiative about the nation's achievements. In all cases, I suggest, scientists on stamps may be termed heroes of science.

How does one define a hero?

Heroes are recognised by society because they bring credit to a country. Celebration of their achievements might occur through national honours, representation in paintings and statues, biographies, inclusion within histories, fellowships and recognition in their own particular spheres of society, and use of their images on coins and banknotes. Within this study, I examine the representation of science and scientists on postage stamps. I have no control over who appears, as it is the issuing postal authority who selects the scientist in order to convey a message. What I shall do is reflect on those choices and the way that science is represented through the potrayal of these scientific celebrities.

In his book, *Celebrity and Power: Fame in contemporary culture*, Marshall describes a celebrity as an individual who is given a greater presence and a wider scope of activity than the person in the street and details "how power is articulated through the celebrity" (Marshall, 1997, p. ix). He states that:

> Celebrity status operates at the very centre of the culture as it resonates with conceptions of individuality that are the ideological ground of Western culture … Celebrity status also confers on the person a certain discursive power: within society, the celebrity is a voice above others, a voice that is channeled into the media systems as being legitimately significant. (Marshall, 1997, p. x)

Marshall discusses the fact that real celebrity requires a substance, an achievement, skill or embedded significance, without which the celebrity sign is entirely image. In this study, the assumption is made that the scientist celebrated on the postage stamp has done something significant to earn the celebration at the time of the issue.

The heroes that follow are selected from the stamp images available, determined by the definitions above, but with specific additions related to scientists as heroes. My additions include scientists who are:

- Major contributors to new (scientific) knowledge or understanding.
- Major influences upon the understanding of science and science methodology beyond their immediate area of expertise.

- Role models who may have had to overcome obstacles in pursuing their objectives.
- Scientists well known to people outside of their immediate area of expertise.
- Possessors of a name recognisable to the general public.
- Scientists who are known to the postal authorities (who will want to concentrate upon the local identities that help define a country's role in science).

A question of numbers?

A measure of a scientist's heroism and their impact upon the world could be the number of times his or her image has been used on postage stamps. I examined two university websites that purport to list the names and images of scientists shown on postage stamps. The University of Frankfurt has a website with a composite of six albums that name the scientist represented, along with the country and year of the stamp issue. There were 823 stamps in the list entitled "physicists on stamps", and 254 named scientists in total. The top six named scientists, listed in the Table 6.1, comprise nearly one-third of all stamps appearing on this website. It offers an indication of the top heroes of science shown for all countries of the world on postage stamps.

Table 6.1: The popularity of images of scientists on stamps shown on the University of Frankfurt website.

Favourites by number of stamps–Percentage of all stamps listed
1) Albert Einstein (1879–1955) – 149 – 8.2%
2) Marie Curie-Sklodowska (1867–1934) – 44 – 5.3%
3) Alexander Graham Bell (1847–1922) – 43 – 5.2%
4) Galileo Galilei (1564–1642) – 38 – 4.6%
5) Nikolaus Kopernikus (1473–1543) – 26 – 3.2%
6) Pierre and Marie Curie – 21 – 2.4%

Source: http://th.physik.uni.frankfurt.de/~jr/physstamps.html

A similar website is compiled by Jeff Miller, who lists images of mathematicians on postage stamps. On this website, there were 720 stamps with the image of a mathematician, with 127 named subjects. The top six, listed in the Table 6.2, comprise 27% of the total number of stamps in this websote. Again, the top six are names we would expect to see in such a compilation.

Table 6.2: The popularity of scientist images shown on Jeff Miller's website.

Favourites by number of stamps–Percentage of all stamps listed
1) Isaac Newton (1643–1727) – 61 – 8.5%
2) Johannes Kepler (1571–1630) – 46 – 6.4%
3) Avicenna (980–1037) – 39 – 5.4%
4) Albrecht Durer (1471–1528) – 19 – 2.6%
5) L. A. de Bougainville (1729–1811) – 16 – 2.2%
6) Leonhard Euler (1707–1783) –16 – 2.2%

Source: http://jeff560.trpod.com/stamps.html.

Having evaluated these numbers as indicators, my study now seeks to explore more deeply the concept of the localised hero at particular times in history, and the messages that such stamps tell. The ten countries of interest to my study have been examined in order. The examples I have selected to discuss in this chapter were born or reared in the country concerned, or are those whose main achievements were made in the country being considered. The frequency of major heroes occuring on stamps is shown in a table at the end of each country section. Also evaluated is the influence of foreign scientists through their celebration on the postage stamps of each country.

The number of scientists considered in this section will not match exactly the number of stamps showing a scientist. Each scientist from each country is counted once, however many times he or she appears on a stamp, and some stamp images laud more than one individual. Figure 6.1 shows an example of multiple celebrants upon a single stamp, which reflects the achievements of three named scientists. The message conveyed by this stamp is that these veterinarians have contributed to French Veterinary Research. Their images are tied together in the design by three arches, textually described as the cradle of veterinary research represented by three men in Lyon, Alfort and Toulouse. The text is required to explain the context, although the stamp is itself a mirror of history.

Figure 6.1: France, 1951. *French veterinary research*, featuring Edmund Nocard (1850–1903), microbiologist veterinarian, Professor H Bouley, veterinarian, and J. B. A. Chauveau (1827–1917), veterinarian. Gibbons catalogue # 1119.

Source: Author's collection.

Explorers on postage stamps, celebrated as both foreign and local heroes

On a count of pure numbers of stamps celebrating a particular individual we find immediate anomalies (Tables 6.3 and 6.4). It would appear that Captain James Cook is the most celebrated local hero of Australia and New Zealand. But Cook was an Englishman, a British Naval Officer on a scientific voyage to study the 1769 transit of Venus from Tahiti, with the secondary objective of discovery in the Pacific. To the conquering populations of these two countries, he might well be a hero whose reputation is regularly embellished with stamp issues with a nation-building emphasis. Cook is anything but a hero to the indigenous occupiers of the land. We might well ask if it has been the stamps that have made the hero, anticipating that the images act as lenses to the historical achievement or if they are the easy reflections (mirrors) of a hero? Cook has also been celebrated as a discoverer of note by Ireland and the United States, who has marked the 200th anniversary of his visits to Alaska and Hawaii. A sample of the stamps featuring James Cook are shown in Figure 6.34. The top six celebrants on the stamps of Australia and top three celebrants on the stamps of New Zealand are shown in Tables 6.3 and 6.4. Explorers are the profession most celebrated by these two countries, until the 1960s, as a means of nation-building and civic education to an immigrant population perhaps unaware of history. James Cook's achievements were also celebrated strongly to commemorate the bicentenary of Australia in 1988. Cook's key scientific advisors, such as Joseph Banks and Daniel Solander, have been nominated as contributors to Cook's success by both Australia and New Zealand. Another common feature shown in the tables is the fact that there is a local explorer in both lists. William C. Wentworth was educated in England but won his repute through inland exploration of Australia (this achievement has been discussed in Chapter Three, see Figures 3.23 and 3.24).

Table 6.3: The top six scientists celebrated on Australian stamps.

Scientist	Achievement	No. of stamps
Captain James Cook	English explorer, navigator	16
Captain Matthew Flinders	Explorer, navigator (Australian coastline)	5
Robert O'Hara Burke and William John Wills	Inland explorers (North/South routes)	5
William C. Wentworth	Inland explorer and polymath	5
Abel Tasman	Dutch explorer, navigator	3
William Dampier	English explorer, navigator	3

Source: Author's research.

New Zealand born Sir Edmund Hillary is a twentieth-century hero who came to the public's attention when he became one of two climbers to conquer Mount Everest in 1953. He later led the 1955–1958 New Zealand Antarctic Expedition. As shown in Table 6.4, he has been shown on New Zealand stamps eight times.

Table 6.4: The top three scientists celebrated on New Zealand stamps.

Scientist	Achievement	No. of stamps
Captain James Cook	English explorer, navigator	11
Sir Edmund Hillary	Apiarist, explorer, diplomat	8
Abel Tasman	Dutch explorer, navigator	5

Source: Author's research.

A similar situation regarding the celebration of local or foreign scientists as national heros arises in regard to the way in which the recipients of Nobel Prizes are claimed by various countries. The country of birth and their adopted countries where the work was carried out can both be used in order to claim a scientist as a national hero. There are also scientists whose scientific contribution to the world is so well respected that many countries celebrate them. I have characterised these as foreign heroes, discussed at the end of this chapter.

The image representation of the scientist

The number of scientists celebrated on stamps within my study is shown in Table 6.5 and Figure 6.2.

Table 6.5: Scientists appearing on stamps.

Type of image	Number
Recognisable image, generally a portrait	491
Generic figures of scientists	191
Recognisable image and appropriate context	937
No personalised image but an attributable, acknowledged scientific achievement	448
Total	2, 067

Source: Author's research.

These totals of the number of scientists are represented in a pie chart Figure 6.2.

6. Scientists on Stamps

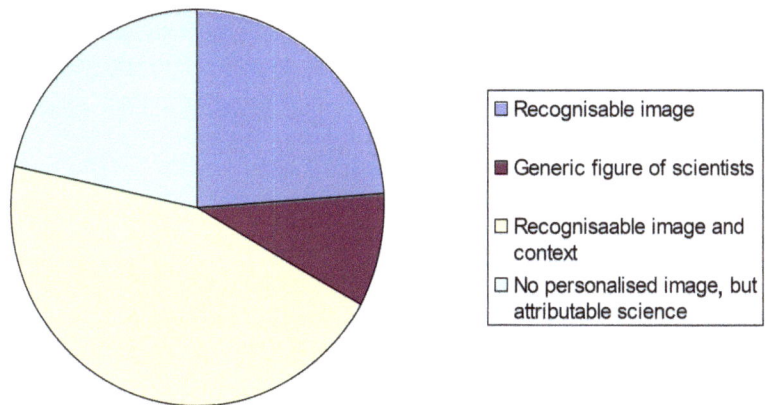

Figure 6.2: Scientists appearing on stamps.
Source: Author's research.

Generic figures do not conform to my definition of a hero of science, as they cannot be named and have a scientific achievement acribed to them. They are included for taxonomic completeness, as extensive use has been made of such images in telling political messages, particularly by the totalitarian states of Eastern Europe, as described in Chapter Four. Only recognisable scientists are included in the remainder of this chapter as heroes.

My study shows that the number of local heroes outnumbers foreign scientists honoured by a country on postage stamps by a factor of 5:1. The ratio of local males to local females is 24:1. A similar ratio exists for foreign heroes. The proportions of local and foreign scientists for each of the ten countries studied is shown in Figure 6.3.

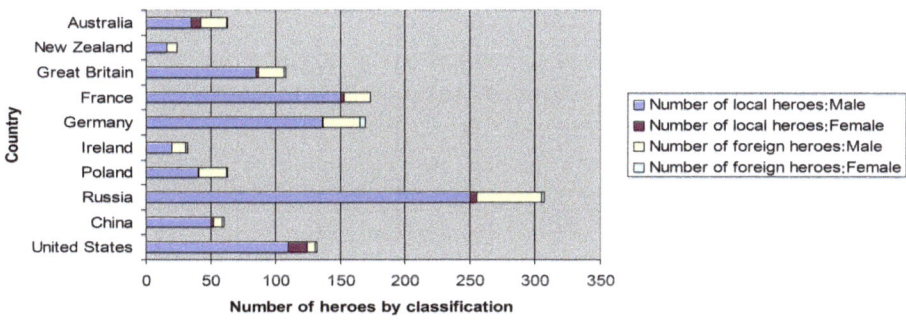

Figure 6.3: The numbers of local and foreign heroes of science appearing on stamps, by gender and country.
Source: Author's research.

Science heroes have been celebrated on stamps in various guises: regularly honoured as heroes in sets that describe multiple achievements, on single stamps celebrating anniversary dates, within institutional celebrations and sets, as celebrities in their own right; or honoured just once or twice, indicating transient fame, even as a scientist.

Local heroes of science

I shall be paying particular attention to two questions in looking at heroes of science:

1. Is the public awareness and perception of science mirrored on postage stamps?
2. Have stamps been issued that contribute to the public awareness of science?

I will examine how the sophistication and detail of context reflects an indication of the public awareness of science by looking at specific examples, following the country by country approach used previously.

Australia

A broad selection of foreign scientists have been represented on Australian stamps: Florence Nightingale (*1955 nursing anniversary*), Alexander Graham Bell (*1976 centenary of the telephone*), Edmond Halley (*1986 return of Halley's Comet*) and the three DNA encoders, Crick, Watson and Wilkins (*2003, 50 years of genetics*) have all been prompted by anniversaries of scientists' discoveries or recognition.

Since the 1990s, Australia has celebrated living scientists on its stamps. This has been a logical development, with Australia Post initiating an annual series in 1997, to coincide with Australia Day, entitled *Australian legends*. At first these legends were all sports people. No celebration of Australian Olympic legends in 1998 would have been complete without recognising a very much active Dawn Fraser. No hue and cry ensued with Dawn Fraser's appearance, and Australia Post have subsequently chose legends without consciously omitting live persons. The 2002 celebrities were five *Legends of medical science* who were acknowledged through portrait photographs, taking two-thirds of the available space. The medical scientists were named, albeit in a very small font. The five 2012 legends celebrated were *Medical specialists*. Fred Hollows and Howard Florey were again honoured within this genre. Five of the seven Australian-born female scientists shown on Australian stamps have been from the three sets celebrating medical prowess issued in 1995, 2002 and 2012.

6. Scientists on Stamps

The first scientist hero celebrated on a stamp who was unequivocally Australian was virologist Sir Frank Macfarlane Burnet. He has been honoured three times on stamps (see Figure 6.4). The first image, from a set of four, was issued in 1975 with the title of *Scientific discovery* and this theme is printed in a very small font on the image. The set is tied together through format. Each stamp has what looks like a paper-tape representation of theme and a large amount of text that is so small that it almost requires a magnifying glass to make sense of. The text reads:

> The foundation of modern *immunology* is the clonal selection theory, which explains how and why anti-bodies are made, occurring in organ transplants. The theory was devised and applied by Sir Frank Macfarlane Burnet in Australia.

The second stamp features Dame Jean Macnamara (1899–1968), the third female named and pictured on an Australian stamp (after Florence Nightingale and Dr Constance Stone). Her discovery, in collaboration with Macfarlane Burnet, of the existence of more than one strain of polio virus was reported in 1931 in the *British Journal of Experimental Pathology* and has been acknowledged as an early step towards the development of the Salk vaccine. The scientific theme of this stamp is shown to be viruses.

The third stamp is from a set of five *Nobel prize winners*, the first time that Australia Post has featured this subject. Four of the five celebrants are scientists. All are illustrated through formal portraits, and the painter's name is shown textually, as is the title of the celebration and the name of the awardee. All items in the set are for the local service fee, suggesting an aspect of civil education and nation-building in the message intended. Over the years, Australia has shown a portrait of the celebrant with some indication of context. Portraits alone are the second-most popular design concept, occuring just behind the classification of portrait with context. For Australia, an example of this trend was shown in Chapter Two, Figure 2.6.

The stamps reproduced in Figure 6.6 appear to challenge the trend towards increasing context. The stamps were, however, issued to send three quite distinct messages and the designers have kept to their briefs, representing McFarlane Burnet's achievements as a contributor to scientific discovery, celebrating medical discovery, and] as a Nobel Prize winner. These stamps were embedded in contexts which were not simply celebrations of an Australian hero.

Figure 6.4: Australia, 1975, 1995 and 2012. *Macfarlane Burnet*. Renniks catalogue # 540 and 1532. The 2012 issue is too new to have been allocated a Renniks catalogue number.

Source: Author's collection.

New Zealand

New Zealand has the smallest number of examples of scientists on stamps of the ten countries included in this study. The numbers were very small until New Zealand Post issued a series of issues to celebrate the millennium. These were largely historical perspectives and all included context as a part of that perspective. I believe that New Zealand Post has recorded the scientific achievements of New Zealanders appropriately and that its millennium issues, discussed in Chapter Seven, make a strong case for science and technology improvements in people's lives. The only foreigners recognised as such and celebrated by New Zealand Post have been explorers, including Neil Armstrong and Buzz Aldrin with the *First man on the moon* issue of 1994, the 25th anniversary of their moon-landing. No recognised female scientist has appeared on a New Zealand stamp.

Ernest Rutherford (1871–1937), has been celebrated three times on New Zealand Post issues. Rutherford was awarded the Nobel Prize in Chemistry 1908 for his work on radioactivity. The first stamp shown in Figure 6.5 celebrates *Rutherford's birth centenary* in 1971. There is an attempt at showing context, with the inclusion of a sketch of the deflection of alpha particles by an atomic nucleus. The second stamps has a portrait incorporated into a background that shows the unmistakable image of the New Zealand flag and an atomic structure. The challenge is declared in the text "leading the way, splitting the atom 1919". The images combine to make the stamp a lens. The third stamp is from a 2008 set of 26 entitled *The A-Z of New Zealand*. Rutherford's role as a hero of science is confirmed within this set: "R is for Rutherford". The three issues reflect the change in emphasis of the image in telling its message. The head of Rutherford dominates the early issue, although there is some attempt at context on the 1 cent stamp, but the representation of the atomic structure is given equal status to the portrait of the young man in the 1999 and 2008 representations. It illustrates a trend that we can recognise for most countries. Prior to the mid-1950s, a portrait was used as the main icon, with or without a textual naming of the scientist. Context was incidental, if it was included at all. From the 1950s, context would be used as the device to explain why the person was being used to tell a message. The later stamp almost sees the head being dominated by the context, in this case, the atomic structure diagram.

Figure 6.5: New Zealand, 1971, 1999 and 2008. Campbell Paterson catalogue # S150, SH105 and S1102.
Source: Author's collection.

Great Britain

Great Britain has, in most cases, highlighted institutions of science in order to celebrate the achievements of its scientists since the mid-1960s. Anniversaries are significant prompts for the issue of sets of, usually, five stamps. Historically, many scientists have gravitated to England to pursue their careers and have their work and achievements absorbed into history. Figures such as Ernest Rutherford (New Zealand), Robert Boyle (Ireland), Gugliemo Marconi (Italy), and Dennis Gabor (Hungary), have been acknowledged as if they were British-born. In this study, I categorise these figures as foreign scientists, as their initial studies were conducted in their birth countries, even though their main consolidated achievements may have been publicised through a Great Britain institution. The only legitimately foreign scientist celebrated by Great Britain is Benjamin Franklin, the American polymath who had appeared twice on British stamps. Franklin was a Fellow of the Royal Society and was chosen by them for inclusion in the issue of 2010 celebrating its 350th anniversary, alongside nine British scientists, (including Boyle and Rutherford), as its face representing 35 years of the Society, 1765–1790. Living persons have not been honoured by Royal Mail except when more modern discoveries or inventions have been featured and it happens that those responsible, although not named individually, are celebrated by implication. James Watson, a decoder of DNA, for example, is still alive.

There are several examples that extending the observation that science and technology messages from Royal Mail use images that show the context of the message but do not necessarily name the responsible scientist. The *British discovery and invention* issue of 1967 featured radar, penicillin, jet engine and television without crediting anyone with discovering these subjects, although later issues featured the same subjects and included the names of Watson-Watt, Fleming, and Whittle. The 2007 *World of invention* set again featured television without naming John Logie Baird, although it is apparent that he was in the designer's mind when he worked on the image. This is stated in the explanatory notes of Royal Mail's review of the postage stamps of that year (Kennedy, 2007, p. 19). One might argue that the British public would know the inventor's names but the stamps, being in a multi-value set, as Royal Mail stamps generally are, means some are destined for use on overseas mail.

There are three quite obvious constraints on Royal Mail's use of a scientist's image to convey a message: the convention to not show living persons on postage stamps, the time taken for the status of a scientist to be realised to convey a science message, and the inbuilt time delays imposed by the lengthy processes of the issuing authority. The monarch and the monarch's family are the only living, recognisable people on the postage stamps of Great Britain. I discuss these issues further in Chapter Eight.

Great Britain has shown Charles Darwin on its stamps 16 times. Naturalist Charles Darwin (1809–1882) was born in 1809. In 1831, he embarked on a five-year survey voyage around the world on HMS Beagle. His studies of specimens around the globe led him to formulate his theory of evolution and his views on the process of natural selection. In 1859, he published *On the Origin of Species*. He died on April 19, 1882, in London. (McCalman, 2009).

Figure 6.6 shows the older Darwin, animal and plant species, and an Indian Ocean atoll anecdotally accepted as being the foundation of his observations, on sets of stamps issued on two life anniversaries. Darwin's portrait is the main feature of each stamp in the early set, appearing along with his signature and two species of the genus of tortoises, iguanas, finches and human skulls, all of which were critical to his evolutionary case. The second set, like the first, provides a lens into Darwin's expertise. Each stamp has the text "Darwin" and the stamps designate and provide the context for the scientist's interests in zoology, ornithology, geology, botany, and anthropology. The stamps are cut in order to fit together as a symbolic jigsaw. The sets are a celebration of the man with an acknowledgement of his theory. The miniature sheet records the survey conducted of the Galapagos Islands conducted by HMS Beagle in 1835 and the species observed by Darwin. The postmark on the sheet records Darwin's reference to the event: "I am turned into a sort of machine for observing facts and grinding out conclusions". This provides an interesting insight into Darwin and the huge amount of labour he invested in his investigations, reflecting his meticulous approach to understanding and proving his concepts prior to publication. All the Darwin images are lenses, encouraging the viewer to become engaged to understand his thesis.

Figure 6.6: Great Britain, 1982 and 2009. *Charles Darwin death centenary* and *Charles Darwin birth bi-centenary*. Gibbons catalogue # 1175–1178 and 2898–2903.

Source: Author's collection.

The first five stamps shown in Figure 6.6 are dominated by Darwin's image. I do need to ask what the casual viewer would make of the five jigsaw images of the second set if they were seen separately, which illustrate the interests of Darwin the generalist naturalist. Certainly the unique shape of the stamps marks them as different, and the theme is shown in text in a bold font. The last five stamps I categorise as showing a scientific image in absentia as it contains no portrait but celebrates a known scientist.

The context trend is apparent in these Darwin examples. We see the man, who will be known to many who see the image, across a range of service fees for future use both in the United Kingdom and worldwide. The first set shows examples from nature that Darwin observed and which prompt edhis theory of evolution. The second set moves into a contextual perspective for all but the portrait on the "within UK" service prepayment. The word "Darwin" does appear in the text, but the image, the context, carries the message.

The British set stamps shown in Figure 6.7 omits any mention of the scientists behind the discovery of DNA; only context is shown. Francis Crick (1916–2004) and James Watson (born 1928) are the featured scientists behind Great Britain *The secret of life* set, which celebrates the progress to decipher the genetic code of the molecule that makes life possible: deoxyribonucleic acid, better known as DNA. The set celebrates the 50th anniversary of the discovery of the structure of DNA by young Cambridge scientists. This issue is definitely a lens, using a humorous approach to prompt interest. However, I learned from Mr Parker,

Head of Stamp Strategy at Royal Mail, that market research had indicated that the humour was not that well received by the general public, and that the approach was not what was expected for a scientific celebration.

James Watson, the American co-discoverer of the DNA helix is a living scientist, a fact at odds with the Royal Mail conventions of not celebrating living persons (other than the Royal Family). This might explain why it is the achievement, featuring a 50% British contribution, does not show the name of the scientists responsible, although the other three stamps in the set issued as *The Scientists' tale* names Faraday, Newton and Darwin. Crick and Watson are in august company. It could also be that the cartoons have been used in this instance to deflect attention away from the discoverers and to prompt thought towards the achievement and its implications.

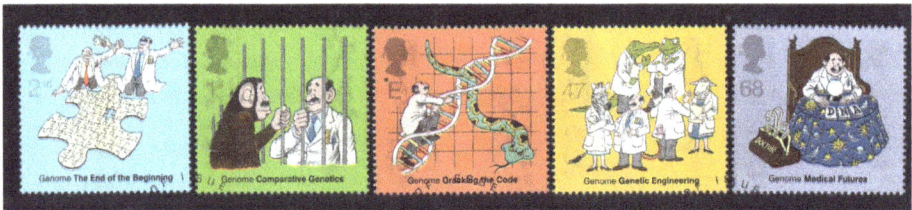

Figure 6.7: Great Britain, 2003. *DNA, the secret of life*. Gibbons Catalogue # 2343–2347.
Source: Author's collection.

France

France has recognised approximately 150 of its scientists in stamps over the 164 years it has been issuing postage stamps. France shows much less consistency in its loyalty to particular scientists than any other country studied (see Table 5.5), although 15 scientists have been used to send a scientific message more than once. Edouard Branly's place in the history of the wireless has been celebrated twice on French stamps, shown earlier in Figure 5.14. Explorers Cartier, D'Urville and La Pérouse have been shown on stamps, as has the French perspective of the history of flight, and the contributions of Le Bris and Dumont are described in Figures 5.27 and 5.32, reflecting France's scientific competences in the nineteenth century. I am unable to explain the decrease in the number of scientists reflected on stamps during the decade 1980–1990, although somewhat fewer stamps were issued overall in this time frame. The percentage of stamps including a picture of the scientist and context is high.

The French Post Office has celebrated French scientists since the 1930s. Its issuing policy appears to concentrate upon the scientist and his or her achievement, and has included context, where appropriate and capable of illustration, in addition to a portraits commemorating its heroes. France has been consistent in its issue of Red Cross Fund charity stamps at a premium and has portrayed many medical scientists. Almost twice as many scientists have been shown on the stamps of France as Great Britain. France seems to place less reliance upon anniversaries for the prompt to issue a new stamp. It is an explorer, Jacques Cartier, the discoverer of Quebec, who has had the most stamps issued in his honour, and this number is only four. France's heritage with respect to pioneers of flight is well recorded on stamps, with Clement Ader and Henri Fabre shown on three stamps each.

The adopted daughter of France, Marie Curie (1867–1934), physicist and dual Nobel Prize laureate, and her scientific relationship with husband Pierre Curie (1859–1906) have been recognised four times by French Poste. These stamps are shown in Figure 6.8. In 1938, the celebration was on behalf of the International Anti-Cancer Fund on the occasion of the 40th anniversary of discovery of radium. 60 years later, 1998, the centenary of the discovery of radium by Marie and Pierre Curie was marked without the use of their images. It is the achievement that is considered as the key to tell its message, which was combined with the 50th anniversary of ZOE Reactor, Chatillon. A complete understanding of the message will require a lot of prior knowledge from the observer. The middle stamp is described in the Gibbons catalogue as Marie Curie and Pitchblende, an amorphous, black, pitchy form of the crystalline uranium oxide mineral uraninite. The final stamp, issued in 2011 to mark the International Year of Chemistry shows Marie Curie in the laboratory with a photographic image that has also been used by the Irish and Polish post offices.

Figure 6.8: France, 1938, 1998 and 2011. *(Pierre and) Marie Curie.*
Gibbons catalogue # 617, 1765 and 3550. WNS catalogue # FR017.11.
Source: Author's collection.

One other heroic French scientist is mentioned here, as the images clearly reflect the introduction of context as a part of the message sent by the issue (Figure 6.9). Louis Braille (1809–1852), was 15 when he developed an ingenious system of reading and writing by means of raised dots. Two years later he adapted his method to musical notation. The first stamp, issued as a single in 1948, a charity stamp with a 4f premium on the service value of 6f, implies that Braille needed no textual description of his celebrity. The second stamp, which appeared 60 years later, includes his name and an image of the awl with which he developed the Braille system of reading. The first stamp is a mirror, the second requires an understanding of the technique Braille invented and is, therefore, a lens.

Figure 6.9: France, 1948 and 2009. *Louis Braille birth bi-centenary commemoration.* **Gibbons catalogue #1023 and WNS # FR001.09.**
Source: Author's collection.

Germany

I have examined the stamps from the five appropriate German postal authorities to see if I could determine any significant favouritism in claiming a scientist for East or West Germany during the 42 years of partition. I could not, although the West Germany post office did issue specific stamps celebrating famous West Berliners.

Analysing the stamps celebrating science, I can discern an increasing level of relevant contextual information with time, but I believe it is true to say that Germany has looked to provide context in its earlier stamps, more so than other authorities. One reason might be that there was an early trend to focus upon the development of flight, which lent itself to meaningful images to accompany portraits of the designer. East Germany, as we have seen, followed Russia's lead in cataloguing Russia's space programme with its inherent visual opportunities.

Physicist Otto von Guericke (1602–1686), best known for his studies of air pressure, is the first (pure) scientist acknowledged with a 1936 German stamp celebration. From left to right, the stamps in Figure 6.10 are: 1936 issue showing a portrait within a frame; 1977 issue, with half of the stamp taken up with the image of von Guericke, with context introduced through a copy of the engraving by Caspar Scott of the Magdeburg hemispheres experiment; 2002 issue illustrating the proof of the power of a vacuum, with 16 horses were unable to separate vacuum-sealed hemispheres. The background to this final image incorporates the Copernicus diagram of the planets. The sequence of images moves from mirror to lens as it moves from the simple portrait to a contextual representation. The third stamp is a good example of a context only image with an eye-catching design and the name of the responsible scientist just legible as the textual explanation of the celebration.

Figure 6.10: Germany, 1936. *Otto von Guericke, 250th death anniversary*; **DDR, 1977.** *Otto von Guericke* from a set of four, *German celebrities*; **Germany, 2002.** *Otto von Guericke, 400th birth anniversary*. **Gibbons catalogue # 605, E 1917 and 3138.**
Source: Author's collection.

The Germanys have celebrated physicist Max Planck (1858-1947), who was awarded the 1918 Nobel Prize in Physics for his discovery of energy quanta, seven times on postage stamps. Four are reproduced in Figure 6.11 and illustrate the development of how images have shown more of the celebrity and his/her achievement over time. The first example is from the set of definitive stamps issued by East Germany in its first year of existence. The set showed ten celebrities with the theme of the *250th anniversary of the Academy of Sciences*. The postal authority is indicating, under the guidance of Russia, that science will be important in the future and that it will be based upon the known institutions and the reputations and achievements of German scientists. West Germany reclaimed Max Planck two years later with a larger portrait from, again, a set of ten *Famous Berliners*. The text clearly names the scientist and the West German part of Berlin as the issuing authority. The third example shows radiation from a black body and the Max Planck quantum theory formula. This stamp is one of two issued under the auspices of *Europa*, the European Postal Authority, which annually recommends a theme to its members. This stamp has a direct reference to the scientist responsible, so it must be assumed that the discovery and its relevance will be known to the general public. Certainly this stamp is a lens, prompting a further look for explanation to the interested viewer. The fourth stamp combines elements of the previous three examples to celebrate both the man—via the photograph, the noting of his birth dates and his signature—and his achievement, shown through his formula. This stamp marks the 150th birth anniversary of Max Planck and was issued as a single item. It is part mirror and part lens.

Figure 6.11: DDR, 1950. Max Planck, *250th anniversary of the Academy of Sciences*; West Berlin, 1952. Max Planck from *Famous Berliners*; Germany, 1994. *Europa discoveries: Max Planck's Quantum Theory*; Germany, 2008. *Max Planck, 150th birth anniversary*. Gibbons catalogue # E25, B99, 2575 and 3530.

Source: Author's collection.

Ireland

Ireland has the second smallest number of scientists on its stamps in the ten countries under examination. Ireland was under British control until the Irish Free State was declared in 1922. Many Irish scientists prior that date were seen to be British and were absorbed into British history and stamp issue, but Ireland has been able to recognise a few heroes. There has been a spike in the number of science issues, reflecting the millennium issues of six stamps. All of the 11 foreign scientists Ireland has celebrated were recognised on these stamps. They are interesting as they recreate, by photograph or painting, moments in time that are mirrors of what occurred and are seemingly without political motive.

The only Irishman to have been awarded a Nobel Science Prize is E. T. S. Walton (1903–1995). The award celebrates his work with John Cockcroft, "atom-smashing" experiments done at Cambridge University in the early 1930s. Figure 6.12 shows a stamp issued on the occasion of his birth centenary. The background shows some of the wording on the Nobel Prize certificate. The stamp was issued as a mirror, it does not confront. Apart from the text, which mentions Walton as a Nobel Laureate in Physics, there is no indication that he is a scientist. The service value is for within Ireland, a suitable celebration of his understated achievement.

6. Scientists on Stamps

Figure 6.12: Eire, 2003. *Ernest Walton, birth centenary*. Hibernian catalogue # C1246.
Source: Author's collection.

Soviet Russia and the Russian Federation

The scientist who has appeared on more Russian stamps than any other is Yuri Gagarin, who has been celebrated 46 times on the stamps of Russia alone. Some examples are shown in Figure 6.18. Two Russian scientists have been celebrated 17 times: Mikhail V. Lomonosov (1711–1765), a polymath who was the first scientist celebrated by the Russian State a year after the creation of the USSR in 1923; and Aleksandr Popov, an inventor of radio who has been discussed earlier in this chapter.

The USSR was formally dissolved on 26 December 1991, after unrest within the constituent republics from 1985. The Russian Federation became the successor state of the Russian Soviet Federative Socialist Republic and is recognised as the continuing legal personality of the All-Union state. The change in constitution, as was shown in Figure 4.29, saw a drop in the number of scientists being recognised on Russian stamp, although the trend from 2000 sees an increasing number of issues. The effect is also discussed later in this chapter.

The point at which context was included in Russian stamps is less well defined than has been previously discussed, as context has been used in many scientific messages from quite early on. This is interesting in itself, but the paradox is evident when Russia continues to use a simple poster-style representation alongside its most contextually rich images. Figure 6.13 illustrates two stamps, both issued in June 1983, to illustrate this observation.

The Representation of Science and Scientists on Postage Stamps

Figure 6.13: Russia, 1983. *20th anniversary of first woman cosmonaut Valentina Tereshkova's space flight and 85th birth celebration of Petr Nikolaevich Pospelov* (1888-1979), Academician and scientist. Gibbons catalogue # 5336 and 5337.
Source: Author's collection.

Dmitri I. Mendeleev (1834–1907), chemist and inventor is an obvious Russian hero of science who has been regularly recognised on postage stamps. Nine stamps have been issued by Russia. In 1934, his birth anniversary was celebrated with a set of four stamps. Three stamps in the set show Mendeleev within a cluttered frame, but one shows him sitting in a high-backed chair in front of a large wall poster containing the data from his periodic table of elements. The stamps shown below in Figure 6.14 celebrate Mendeleev as, left to right: One of the 1951 set of *Russian Scientists*. The facial image is complemented with an image of a chemical retort for context. The *50th death anniversary* issue follows the full-face poster style within a complicated framing and description of the event being marked. The top border shows a monumental building that I cannot recognise, although the Russia catalogue notes the intention for this stamp to be reissued with an overprint notifying the general public of the VIII Mendeleev Congress 1958 (Stanley Gibbons Publications, 2010, Part 10, p. 74). The third stamp again uses a portrait to dominate the image and reinforce the perception of the scientist as an older man, deep in thought at his deck as he ponders the periodic table. The context is well thought out, with Mendeleev holding a pencil and some of his workings shown with chemical laboratory equipment in the immediate foreground. This stamp celebrates the *centenary of the Periodic Law* in 1969. At the same time, the miniature sheet reproduced in Figure 6.14 (at a small size) was published. The embedded stamp is a simple portrait, but the complete design of the sheet is used to expand upon the periodic table and shows a signed and dated work paper. The sheet is garnished with a laurel leaf motif, denoting an honour to celebrant Mendeleev. The 6K stamp and the 30K sheet are both lenses, invoking a curiosity to understand the achievement.

6. Scientists on Stamps

Figure 6.14: Russia, 1951, 1957 and 1969. Dmitri Mendeleev, from a set of 16 *Russian scientists*; Dmitri Mendeleev, *50th death anniversary*; and *Centenary of Dmitri Mendeleev's Periodic Law*. Gibbons catalogue # 1720, 2048, 3696 and MS3697.

Source: Author's collection.

Mathematician Mstislav Vsevolodovich Keldysh (1911–1978), was the President of the Academy of Sciences of USSR from 1961–1975. In addition to the Soviet scientific research ship named Akademik Mstislav Keldysh, a crater on the moon and a minor planet (2186), have been named in his honour. Among the scientific circles of the USSR, Keldysh was known by the epithet the Chief Theoretician and was a key figure behind the Soviet space programme (Sergey and Kazbek, 2013). As Novokic puts it:

> Mstislav Keldysh, was a very talented mathematician in the theory of functions of a complex variable and in differential equations. An especially fundamental contribution was made by him to applied branches of aerodynamics. He was a Chief Theoretician-adviser of the government and an organizer of computational work related to jets and space in 1940–1960s. He was a widely known person in the Soviet society. All information on the work of such people was classified and not reflected in the world press (Novikov, 2012).

The Representation of Science and Scientists on Postage Stamps

The first time we meet mathematician Keldysh on a postage stamp (Figure 6.15), his profile, name and life-dates cover one quarter of the available surface space, with the dominant image being the Soviet scientific research ship named after him. A year later, his image is shown on a poster-style acknowledgement of his 70th birth anniversary. It is another 30 years before the birth centenary celebration context confirms that he is a mathematician whose work has had an impact on the design of space satellites. A mathematical formula is clearly stated. The fact that, (as confirmed by Novokic, 2012), an unheralded scientist has a vessel named after him illustrates the importance that Russia placed upon scientific endeavour during the Soviet period. Of the three messages, the first two are mirrors but the additional data integrated into the third makes it a lens.

Figure 6.15: Russia, 1980, 1981 and 2011. Mstislav Keldysh, *Soviet Scientific Research Ships*, (2nd series); Mstislav Keldysh, *70th birth anniversary*; and Mstislav Keldysh, *Birth centenary*. Gibbons catalogue # 5058, 5091. WNS catalogue # RU001.11.

Source: Author's collection.

Russia has been quite generous in its celebration of foreign scientists, in images used to convey positive messages of scientific progress. A significant number of these scientists have actually been foreign cosmonauts who were part of the continuing Russian space missions of the 1970s and 1980s. Russia sought political advantage in carrying representative technicians from other countries and optimised publicity with the issue of commemorative stamps. Having decided to include space research in my study, I would be remiss not to highlight the fact that Yuri Gagarin (1934–1968), the first man in space, is undoubtedly a hero of Russia. He has also appeared on the stamps of France, Germany and Poland. In 2011, the smiling face of Gagarin in a space-suit helmet was shown to mark the 50th anniversary of his flight. Figure 6.16 reproduces an earlier issue of four stamps celebrating the (forever young) cosmonaut.

Figure 6.16: Russia, 1991. *Yuri Gagarin, 30th anniversary of the first man in space*. Gibbons catalogue # 6238-6241.

Source: Author's collection.

In 1987, The Russian postal authority looked outside of its own confines and issued the set of three stamps shown in Figure 6.17. The three scientists recognised are shown in a non-political way. From left to right:

- Ulugh Beg (1394–1449), Timurid ruler as well as astronomer and mathematician, on a stamp celebrating the *550th anniversary of New Astronomical Tables*.
- Sir Isaac Newton (1643–1727), English physicist, mathematician, astronomer and philosopher, celebrated on the *300th anniversary of Principia Mathematica*.
- Marie Skłodowska Curie (1867–1934), Polish physicist, chemist and Nobel laureate, celebrated on her *120th birth anniversary*.

Each stamp was issued with a se-tenant label bearing an inscription detailing the subject's scientific achievements. As they were issued at the domestic service fee, and the descriptions are in Russian, it can be assumed these were issued as lenses to notify the general public of these particular anniversaries as a civic education message.

Figure 6.17: Russia, 1987. *Scientists' anniversaries*. Gibbons catalogue # 5801–5803.
Source: Author's collection.

China

It is more difficult to mark out a Chinese hero of science based upon frequency of celebration on stamps. Zeng Hi (1371–1433), the Hui-Chinese mariner and explorer, has been shown on two sets, seven stamps, and Xu Xiake (1587–1641), another mariner, on just one set of three stamps. These are the only persons so celebrated. China has made extensive use of generic figures on its stamps illustrating government policy supporting science by providing images of workers being assisted by the benefits of science and technology. The Chinese Post Office has recognised its Chinese-born scientists through two series of issues. These have been titled *Scientists of Ancient China*, three sets featuring 12 people, and *Scientists of modern China,* five sets (between 1988 and 2011) featuring 20 people. These two series constitute 60% of all the local scientists celebrated on stamps, and there has been no duplication of celebrant. Context has been a feature of the middle set of the ancient scientists, with two stamps issued for each celebrant, one showing the person and the other his achievement in context. The modern issues also show context within the design. Examples of these two series are shown in Figures 4.11, 4.12, 6.18 and 6.19.

6. Scientists on Stamps

Figure 6.18: China, 2002. *Early Chinese Scientists*. Bian Que (C500BC), earliest physician, Liu Hui (3rd century) mathematician, Su Song (1020–1101), astronomer, and Song Yingxing (1587–1666), scientist. Gibbons catalogue # 4747–4750.

Source: Author's collection.

Figure 6.19: China, 2006. *Modern Scientists*, Liang Xi (1883–1958), forester, Mao Yisheng (1896–1989), civil engineer, Yan Yici (1900–1996), physicist, and Zhou Pelyan (1902–1993), physicist. Gibbons catalogue # 5086–5089.

Source: Author's collection.

Generic figures optimising the benefits of science and technology in a country politically endorsing science has been the norm for China. These images peaked in the 1960s and 1970s. China Post's more modern designs follow the western

United States

The United States was an early adopter of the postage stamp in 1847 as a fiscal device, and was seemingly unconstrained by its history in needing to develop nationalistic icons. The Founding Fathers' portraits were used until 1869, at which time specific events, later politicians and the American eagle, flags and shields began to appear. These early issues carried no text other than the country name and the value of the service. Text was introduced from the 1930s, introducing the link to context. The United States Postal Service issuing policy relating to the actual number of stamps issued is not dissimilar to that of Australia and New Zealand. 38 stamps have celebrated Christopher Columbus, the earliest appearing in 1869, with the next in 1893. Early twentieth-century issues have featured land explorers, some of whom were not locally born.

The United States Postal Service rigidly adheres to its policy not to feature a living person on its postage stamps, and it has been emphasised in recent press articles that there is little chance of this policy being changed, although the Post Master General has said he endorses the change. The United States Postal Service is also conservative when it comes to honouring its scientists, preferring to wait for the scientist's place in history to be fully recognised before celebrating them on a postage stamp. The United States is different to many issuing authorities in that it regularly issues its definitive stamps with issues named *Prominent Americans*, for example, and uses a simple named portrait as the main image. The Jonas Salk stamp shown in Figure 6.21 is a good example

The postage stamp is used to tell a wide variety of messages. Within this study I have looked at the achievements of The Wright Brothers, Thomas Edison, John James Audubon, Alexander Graham Bell, Einstein, the Antarctic explorers, and the first man on the moon. More than 100 American scientists have been shown on US stamps, many as *Distinguished Americans* within sets of definitive stamps. Recently the United States Postal Service has begun to issue sets of four scientists. Three scientists were featured within the millennium issues of the US and allow for the context examination to continue, as these stamps carried a textual explanation on the back of the stamp to actually elucidate the message being told. Two of these are shown in Figures 6.20 and 6.21.

John F. Glenn (born 1921), was the first US astronaut to orbit the earth. In the winter of 1962, the US needed a hero. Americans had yet to recover from the

Soviet Union's launch of the first spacecraft, Sputnik, in October 1957, a rude jolt to confidence of the world leaders in all things technological. The space race was on.

> The US lagged, managing only two 15-minute suborbital astronaut flights—only five minutes of weightlessness each time. Then, on Feb. 20, 1962 a Marine Corps fighter pilot from small-town America stepped forward. The astronaut was John Glenn squeezed into the cockpit of a Mercury spacecraft called Friendship 7, launched by an Atlas rocket from Cape Canaveral, Florida Glenn circled the Earth three times, becoming the first American to orbit the planet (Wilford, 2012).

Two stamps have been issued to celebrate the achievement (Figure 6.20). You will note that neither shows the name of John Glenn within its image, although the 2000 issue does name him on the reverse, explaining his return to space in the 1990s.

Figure 6.20: United States, 1962 and 2000. *Project Mercury, first orbital flight by US astronaut* and *Celebrate the century, the 1990s*. Scott catalogue # 1193 and MS 3191.
Source: Author's collection.

American physician and epidemiologist Jonas Salk (1914–1995) developed the first effective vaccine against poliomyelitis, a crippling disease that killed more than 3,000 Americans at its peak in 1952, and leaving many thousands more crippled or paralysed. The two Salk stamps shown in Figure 6.21 celebrate this achievement. The first shows a doctor administering an injection to a girl, with text on the back of the stamp offering contextual explanation and explaining its historical and medical importance. The second stamp was issued as one of a definitive set in 2006, a a prompt for Salk's selection was probably the 50th anniversary of federal approval for the use of the Salk polio vaccine.

The Representation of Science and Scientists on Postage Stamps

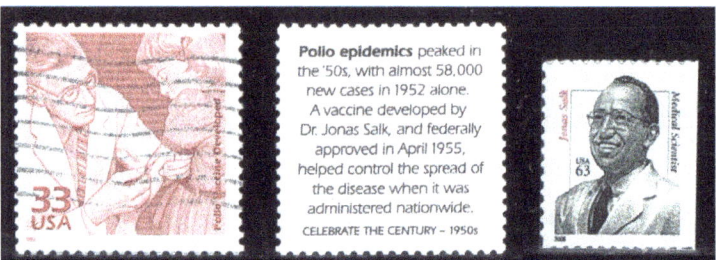

Figure 6.21: United States, 2006. *Jonas Salk* from *Celebrate the century, the 1950s,* and *Jonas Salk* from the series, *Distinguished Americans*. Scott catalogue # MS 3187. WNS catalogue # US033.06.
Source: Author's collection.

Two US women have been awarded the Nobel Prize for Physiology or Medicine, and both have been proclaimed on United State Postal Service issues. Polish-born Gerti Cori (1896–1957) was and jointly awarded in 1947 with her husband, Carl Cori, and B. Houssey for the discovery of the course of the catalytic conversion of glycogen. Barbara McClintock (1902–1992) won the 1983 Prize for her discovery of mobile genetic elements. They were celebrated within the first two *American Scientists* series. The two stamps are shown as Figure 6.22 and are, on the surface, mirrors, with equal prominence given to the scientist and the context within the design.

I describe the first stamp shown in Figure 6.23 as perhaps a mirror, but also a lens. This 41 cent stamp was reported by the Associated Press to have a printing error in the chemical formula for glucose-1-phosphate. The stamp was distributed despite the error being known (foxnews.com, 2008). An error delights the philatelist, and this error was pursued at a fervent level by scientists and philatelists in the editorial columns of national dailies and stamp magazines, out of which discussion came the correct formula (Figure 6.22).

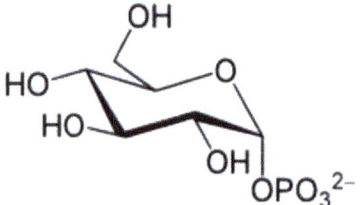

Figure 6.22: Correct formula for glucose-1-phosphate, which was shown incorrectly on the US postage stamp.
Source: Author's research.

202

6. Scientists on Stamps

Figure 6.23: United States, 2008 and 2005. *American Scientists*, series two and one, *Gerti Cory* and *Barbara McClintock*. WNS catalogue # US004.05 and Scott catalogue # 3906.
Source: Author's collection.

The US has issued almost the same number of stamps featuring females as the other nine countries in my study added together. Of the American female Nobel Prize winners in the scientific categories—one for physics and two for physiology or medicine—two have appeared on stamps shown in Figure 6.23. Jane Addams, the 1931 Nobel Peace Prize recipient, although a social activist, was included in the 1940 issue of *American scientists*. The stamp is reproduced in Figure 3.31.

Summary of the representation of the scientist on stamps

How the scientist is represented varies country by country, dependent upon the country's style and convention guide, policy, political motives and culture. As shown in Table 6.3 and Figure 6.2, the science hero might be represented by a portrait, portrait and context, or via a scientific achievement through which the scientist is known well enough to not show the portrait, although he or she might also be named. Figure 6.24 demonstrates how the representation changes, for example, between France and Germany, two European countries. Until the mid-1990s, France favoured a representation showing the scientist's portrait, which has diminished, whereas Great Britain, who have only shown celebrities (other than royalty) on stamps since the mid-1960s, have represented the science of known scientists without showing the portrait.

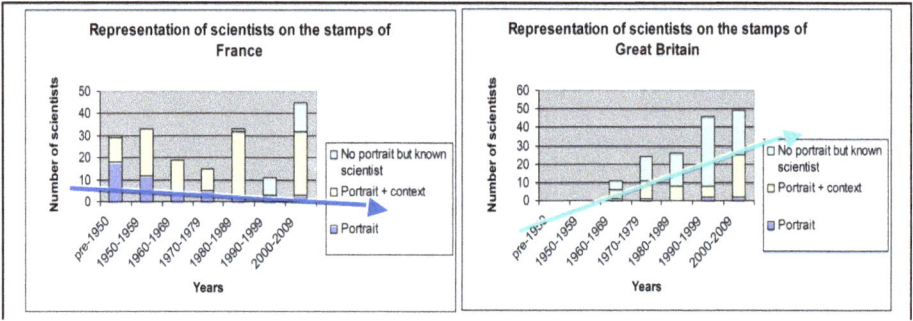

Figure 6.24: The representation of scientists on the stamps of France and Great Britain. NB: the scale of the y-axis of the charts differs.
Source: Author's research.

Looking at the two Germany results shown in Figure 6.25, generic figures have mainly been used to illustrate the good that will accrue from the implementation of science within the East German planning regimes. Generic figures minimise the cult of the individual and emphasis the pre-eminence of the State in people's lives. The high level of portrait with context occuring from the 1960s for East Germany is the result of celebration of space research events and anniversaries of Russia. The trend for West Germany has been a move from the scientist's portrait appearing alone to, in later years, portraits appearing in association with a contextual element. As the arrows show, both ideologies moved from celebrating the scientist through a simple portrait to a portrait including context.

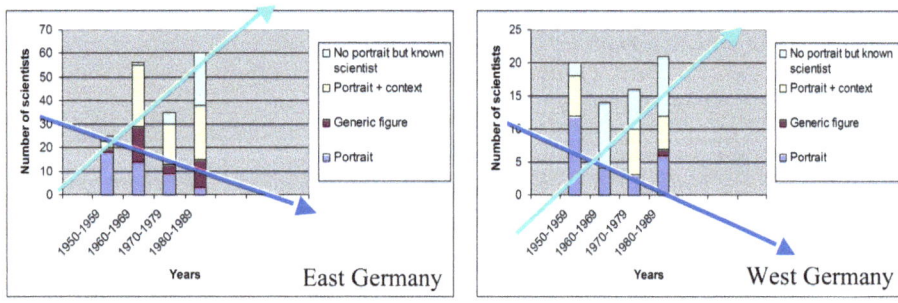

Figure 6.25: Classifications of the images of scientists on the stamps of East and West Germany. NB: the scale of the y-axis of the charts differs.
Source: Author's research.

The number of Russian issued stamps declined after the death of Stalin. Similarly, the number of Chinese issued stamps declined after the death of Mao Zedong, indicating that leadership change can have a strong influence. At the end of 1991, the USSR broke up. This study has examined the Russian Federation stamps as a continuation of the USSR. The number of stamps issued has fallen dramatically with this constitutional change. Figure 6.26 illustrates the change when reviewing the number of scientists being shown on stamps. Brunn has examined stamps as messengers of the USSR's political transition to the Russian Federation, confirming my observations (Brunn, 2011, pp. 19–36).

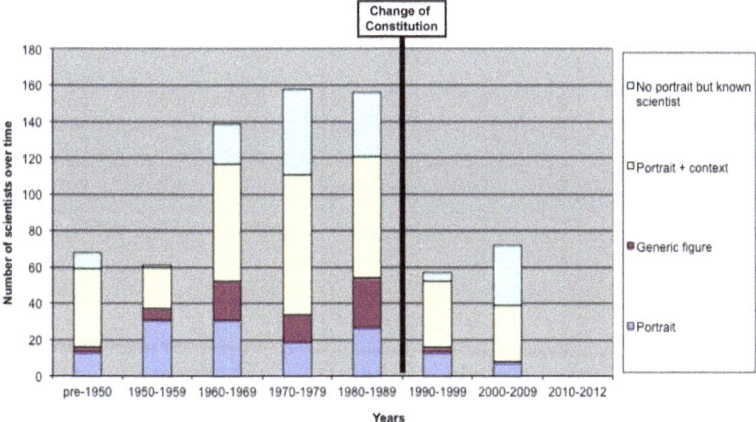

Figure 6.26: Constitutional change and the number of scientists featured on Russia's stamps.
Source: Author's research.

Russia represents one-third of the stamps studied here, and the dramatic drop in the number of stamps and the number of scientists featured on their stamps after the break-up of the USSR has an effect on the graphical representations that show how the importance of context has developed. I describe the effect as splitting into two phases after the end of 1991. Generic figures have been removed from Figure 6.27.

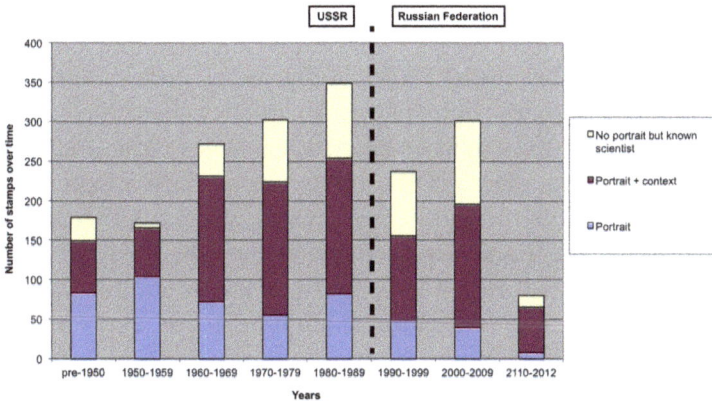

Figure 6.27: The number of scientists represented on stamps over time, showing a trend to more context within the stamp image and the effect of the break-down of the USSR upon the number of issued stamps.
Source: Author's research.

In Figure 6.27, arrows indicate the trends to show that, within the ten countries of the study: the representation of the scientist by his portrait alone is declining, the levels of context in association with a portrait in telling the stamp's message is increasing, as is the representation of the scientific achievement of known scientists without using their portrait.

The breakup of the Soviet Union has significantly reduced the number of science stamps issued by that postal authority, especially when viewed over a decade. Figure 6.28 reflects the change this makes to the total count of scientists on stamps.

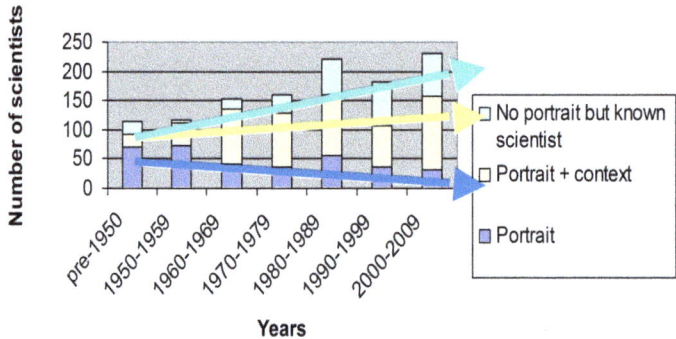

Figure 6.28: The contextual representation of all scientists examined in this study, omitting the stamps of the USSR and the Russian Federation.
Source: Author's research.

Removing Russia from the analysis, the trends remain similar, if less marked. The use of portraits of scientists occurs less over time while there is a continuing increase in showing context.

Moving from a style that shows heroes of science by portrait alone, the stamp designer now incorporates context in order to strengthen the message being told. In general terms, this might be described as a move from a mirror to a lens in order to tell a scientific message, illustrating the move from expecting a public understanding of science to a public awareness of science. The sophistication of the context to enhance the message being shown on the stamp has consistently risen. Context allows the designer an opportunity to portray ideas requiring attention, which have been examined as lenses. Science heroes are recognised and celebrated through these mechanisms, although on later stamps the hero might not be shown, or even mentioned, in addition to their achievement.

International heroes of science

This section focuses upon those heroes whose status and achievements have been celebrated not only in their home country, but also by other countries to whom they are foreigners. Table 6.6 outlines the data.

It is understandable that any particular country will want to publicise the scientific achievements of its own citizens. These achievements are a source of national pride. Publication serves the many roles of the postage stamp in sending a message that has a historical, cultural, philosophical or political motive. The stamp's message will certainly be seen by the country's public during the course of delivery to its final destination. Messages for internal consumption may be mirrors that reflect reality and provide a historical or cultural narrative to promote nation-building and civic education. Stamps that are lenses are intended to promote a change in perception and or behaviour; those issued with a philosophical and political motive will, in the main, be directed towards internal action. So why would a country promote foreign scientists as message bearers?

Some arguments will be developed in the next chapter, in an examination of the use of scientific images at two specific points in time. I suggest here that representations of foreign scientists who are recognised to have contributed significantly to the public good are used to acknowledge their contribution as a positive factor worth the message. Acknowledging the contribution of Marie Curie, for example, hints at a gender balance, even if one might not exist. A country such as China, which has had a reputation for being insular, has looked outside its boundaries to honour: Copernicus (1953, 480th birth anniversary), two foreign doctors who contributed in the second Sino-Japanese War (1960 and 1982), Albert Einstein (1979, birth centenary), Robert Koch (1982, centenary of the discovery of the cause of tuberculosis), Edmund Halley (1986 appearance of Halley's Comet) and Christian Golbach (1999, 50th anniversary of the Chinese Academy of Science and the Goldbach conjecture).

"The first man on the moon", is generally shown as a generic figure, although most people know the names of Neil Armstrong and Edwin (Buzz) Aldrin. Appearing at the same level of recognition is Charles Darwin.

Table 6.6: The scientists who have been celebrated both in their country of origin and as a foreign scientist.

	Aus	NZ	GB	Fr	Ge	Irl	Po	Ru	Ch	US	Scientist Total
Albert Einstein				✓	✓	✓	✓	✓	✓	✓	7
Copernicus				✓	✓		✓	✓	✓	✓	6
Edmond Halley	✓		✓		✓		✓	✓	✓		6
James Cook	✓	✓	✓			✓				✓	5
Robert Koch				✓	✓		✓	✓	✓		5
"The first men on the moon"		✓		✓		✓	✓			✓	5
Charles Darwin			✓		✓	✓	✓	✓			5
Alexander Bell	✓		✓		✓					✓	4
Marconi			✓		✓	✓				✓	4
Columbus		✓					✓	✓		✓	4
Isaac Newton			✓		✓		✓	✓			4
Marie Curie				✓		✓	✓	✓			4
Louis Braille				✓		✓	✓	✓			4
Yuri Gagarin				✓	✓		✓	✓			4
Avicenna				✓	✓		✓	✓			4
Florence Nightingale	✓		✓		✓						3
Benjamin Franklin			✓					✓		✓	3
Lord Rutherford		✓	✓					✓			3
Pierre Curie				✓			✓	✓			3
Albert Schweitzer				✓	✓		✓				3
Country Total	4	4	9	9	11	8	14	14	4	8	

Source: Author's research.

It would appear that the European countries are more likely to celebrate the achievements of foreign scientists, perhaps because of their long scientific history. It would seem that Australia, New Zealand and Great Britain have a common isolation that has not recognised Einstein, Copernicus, Koch, Marie Curie, Braille, Yuri Gagarin, Avicenna, Pierre Curie and Albert Schweitzer on their stamps.

I shall review these international heroes here.

6. Scientists on Stamps

Albert Einstein

Figure 6.29: Albert Einstein (1879–1955).
Source: Author's collection.

I had anticipated that Einstein would win any poll of internationally recognised scientists. What is interesting is that he and his work have not been shown on the stamps of Great Britain, Australia and New Zealand. Seven countries have used Einstein to tell a message. Ireland have used his image twice, the first time as a significant millennium subject in which Einstein, the man, is celebrated, as one of six stamps reflecting world historical discoveries, with no textual elaboration apart from his name and life-dates (thus a mirror of reality), and the second time as one of a set of three stamps celebrating 2005 as the International Year of Physics.

Apart from the East German 1979 issue and the 2001 Irish stamp from the *Millennium, discoveries* series, Einstein is always shown as an older man. In 2005, he was chosen as the Face of the United Nations' Year of Physics. Three European examples celebrate the 2005 Year of Physics. Six of the ten stamps shown include Einstein's formula for energy, confirmation that this is arguably the most well-known formula in the world. I am unable to explain the pixelation of the background to the French stamp unless the designer is examining the photo-electric effect that is illustrated on the West German acclamation. The use of Einstein's image on the East German stamp superimposed over a sketch of a Soyuz space craft, part of the 1978 Soviet-East German Space Flight issue, and the symbolic flight badge, elevate Einstein's science contribution into space.

I categorise these stamps as mirrors, celebrations of Einstein, acknowledging his influence upon scientific thought and implying the implementation of his ideas into the knowledge of these seven countries.

Nicolaus Copernicus

Figure 6.30: Nicolaus Copernicus (1473–1543).
Source: Author's collection.

The nine stamps reproduced in Figure 6.30 recognise Copernicus's heliocentric model, with the sun at the center of the universe, demonstrating that the observed motions of celestial objects can be explained without putting Earth at rest in the center of the universe. His work stimulated further scientific investigations, becoming a landmark in the history of science that is often referred to as the Copernican Revolution.

There is a continuing controversy as to whether Copernicus was a Pole or a German. Davies writes that Copernicus, as was common in his era, was largely indifferent to nationality, being a local patriot who considered himself a Prussian (Davies, 2005, p. 20). The Polish postal authority has certainly made the most of their hero, issuing 38 stamps to celebrate his achievements. The Polish issues cover Copernicus's life as well as statues, institutions and buildings named in his honour. The first of Copernicus's great successors was Tycho Brahe, though he did not think the earth orbited the sun, followed by Johannes Kepler, who had worked as Tycho's assistant in Prague (Repcheck, 2007, p. 79).

The image used on all of these stamps is based upon the 1580 portrait in Toruń Old Town City Hall. Five of the seven stamps show the circular artifact Copernicus constructed to explain his theories in the *Commentariolus*. The image shown from Poland is from the Matejko painting "Conversations with God". All of the stamps are mirrors celebrating the man.

Edmond Halley

Figure 6.31: Edmond Halley (1656–1742).
Source: Author's collection.

Edmond Halley was an English astronomer, geophysicist, mathematician, meteorologist, and physicist, best known for computing the orbit of the eponymous Halley's Comet. He was the second Astronomer Royal in Britain.

Astronomy and astronomers, it would appear, are popular subjects for postage stamps. It makes sense because the sky is common to mankind and has been important culturally during all of recorded time.

All of the stamps shown in Figure 6.31 have been issued on the occasion of the 1986 appearance of Halley's Comet. It is interesting to compare these stamps with those that have celebrated Copernicus. Halley's portrait is only used on the Russian miniature sheet, although a facial sketch precedes a feathered tail in the first of the British stamps. All of stamps have a representation of the comet and its distinctive tail. The Australian stamp includes a sketch of the comet's orbital path and is both mirror and lens. Confusingly, the main image is the Parkes Radio Telescope, an Australian icon in its own right, although the comet is a purely visual object with no radio emissions.

The British stamps tell a tale through its four images: *Dr Edmund Halley as a comet, Giotto spacecraft approaching comet, Twice in a lifetime* and *Comet orbiting sun* and *planets*. It is a lens to the aspiring astronomer.

During 1984–1986, Russia sent its message about its continuing space research and the voyage of the Vega 1 space probe to Venus, which included a fly-by of Halley's Comet, with three stamps and a miniature sheet,. This series is a lens, telling a visually detailed story.

Captain James Cook

James Cook was born in 1728 in Marton, England. While working in the North Sea, Cook spent his free time learning mathematics and navigation. This led to his appointment as mate. In 1755, he volunteered for the British Royal Navy and took part in the Seven Years War. He was an instrumental part of the surveying of the St. Lawrence River, which helped in the capture of Quebec from the French. Following the war, Cook's skill at navigation and interest in astronomy made him the perfect candidate to lead an expedition, planned by the Royal Society and Royal Navy, to Tahiti to observe the infrequent passage of Venus across the face of the sun. Precise measurements of this event were needed worldwide in order to determine the accurate distance between the earth and sun. After the stop in Tahiti, Cook had orders to explore and claim possessions for Britain. He charted New Zealand and the east coast of Australia, known as New Holland at the time (Rosenberg, 2012).

The first stamp in Figure 6.32 is unique. It comes from the first set of commemorative stamps in the world, the 1888 Centenary of the First Fleet's arrival in Sydney, issued by the Colony of New Sort Wales. Subsequent issues by five countries qualify Cook as a hero of science. It is his image that dominates all the stamps, although three also show his ship, The Endeavour. Australia Post has issued 16 stamps commemorating Cook's voyages of discovery to Australia, although these days it is more politically correct to talk about European Discovery. New Zealand has produced 11 stamps featuring James Cook. The Irish issue celebrates Cook in its *Millennium Series – Discoverers*. The 1978 US set commemorates an American connection, the 200th anniversary of Cook's visits to Hawaii and Alaska. All these stamps celebrate the man and are mirrors, not lenses.

Figure 6.32: Captain James Cook (1728–1779).
Source: Author's collection.

The Representation of Science and Scientists on Postage Stamps

Robert Koch

Figure 6.33: Robert Koch (1843–1910).
Source: Author's collection.

Robert Koch, the son of a mining engineer, astounded his parents at the age of five by telling them that he had, with the aid of the newspapers, taught himself to read, a feat which foreshadowed the intelligence and methodical persistence which were to be so characteristic of him in later life … Some two years after his arrival in Berlin to work at the Imperial Health Bureau Koch discovered the tubercle bacillus and also a method of growing it in pure culture. In 1882 he published his classical work on this bacillus. He was still busy with work on tuberculosis when he was sent, in 1883, to Egypt as Leader of the German Cholera Commission, to investigate an outbreak of cholera in that country. Here he discovered the vibrio that causes cholera and brought back pure cultures of it to Germany. He also studied cholera in India … In 1905 he was awarded the Nobel Prize for Physiology or Medicine. In 1906, he returned to Central Africa to work on the control of human trypanosomiasis, and there he reported that atoxyl is as effective against this disease as quinine is against malaria. Thereafter Koch continued his experimental work on bacteriology and serology (Nobel Foundation, 1967).

Almost the same portrait of Koch has been used for the seven stamps reproduced here, although the beard and hair have been whitened for the Poland issue. Four of the stamps show just the portrait, although in one Koch is reading. These obviously commemorate the man, implying that his deeds are known. Four of the stamps are issued in 1982, the centenary of the discovery of tubercule bacillus.

Three of the stamps show context, with the French and Chinese stamps including a microscope and the image of what Koch might have seen. These three items are lenses. In 2005, we see the German issue showing more clearly defined context, with a clear image from the microscope slide. These stamps do not challenge and are mirrors of reality.

'The first men on the moon': Neil Armstrong and Edwin (Buzz) Aldrin

Figure 6.34: The first men on the moon, 1969.
Source: Author's collection.

On 20 July 1969, Neil Armstrong became the first man to walk on the moon; Buzz Aldrin, following his Apollo 11 crewmate, was the second. Aldrin piloted the lunar module during that mission and spent over two hours on the Sea of Tranquility. Both men were dressed similarly in a pressurised suit that masked their identities behind a visor. This is the image that has become the icon for the event and the overall achievement. The two men have become synonymous with the event, as has Michael Collins, the Command Module pilot. The images used are, however, of the men who walked the moon's surface.

The United States forecast the event through a stamp two months earlier, acclaiming the Apollo 8 mission which put men into orbit around the moon during December 1968, with the image of Earth, the blue planet rising over the surface of the moon with the words: 'In the beginning God …'. The 1969 US stamp included in the figure above was issued in September 1969 as an Air Post stamp, in anticipation of overseas use. The image reproduces the scene of the Apollo 8 stamp, with the image of an astronaut first setting foot on the moon. Russia did not acknowledge American achievements on its postage stamps during the Cold War and the space race, until the 1975 *Apollo-Soyuz* space link project commemoration, which shows the portraits of three NASA astronauts with two USSR cosmonauts.

Eastern Europe followed the Russian example in ignoring the US moon landing. Of the stamps shown here, one marks the 20th anniversary and three celebrate the 25th anniversary of the moon landing. The French and Irish stamps are from their millennium issues, highlighting the significant events of the twentieth century. The New Zealand stamp is a holograph that attracts attention, as it is different. I categorise all the images as mirrors of the event.

Charles Darwin

Figure 6.35: Charles Darwin (1809–1882).

Source: Author's collection.

Charles Darwin was a British scientist who laid the foundations of the theory of evolution and transformed the way we think about the natural world. Great Britain has celebrated Darwin with 16 different stamps, including the unique set of five stamps cut into the shape of a jigsaw puzzle as an analogy of his life's work. The five countries featuring Darwin have all used images of him as an older man.

Darwin's life dates must be on the permanent diary of anniversaries for potential stamp issues. All the stamps shown relate to anniversaries: 150th birth centenary in 1958, 1982 death centenary and 2009 birth bicentenary. There is no doubt

Darwin's influence is acknowledged world-wide. The theory of evolution still causes controversy, but the stamps illustrated are mirrors of acknowledgement except, perhaps, the Great Britain set, which shows different species of similar animals facing inwards to the Darwin image, as these do prompt a further look. I categorise these stamps as lenses.

Avicenna

One eastern scientist features as a hero of science, although I do not know if celebration of his person contributes to the western/eastern science debate. Four European countries have chosen to honour Avicenna, Abū ʿAlī al-Ḥusayn ibn Abd Allāh ibn Sīnā (980–1037), the Muslim physician, most famous and influential of the philosopher-scientists of the Islamic world. He was particularly noted for his contributions in the fields of Aristotelian philosophy and medicine. He composed the Kitāb al-shifāʾ *Book of the Cure*, a vast philosophical and scientific encyclopaedia, and Al-Qanun fi al-Tibb *The Canon of Medicine*, which is among the most famous books in the history of medicine. Avicenna was born in a part of Persia, now known as Uzbekistan, which might explain the 1980 Russian stamp and the earlier marking of the birth millenary of Abū Rayḥān Muḥammad ibn Aḥmad Bīrūnī (973–1048), another Persian Muslim scholar and polymath honoured as the most original polymath the Islamic world had ever known.

Figure 6.36: Avicenna (980-1037).
Source: Author's collection.

Summary of the representation of international heroes on stamps

In most cases, a country tends to mark the achievement of a foreign scientist on the occasion of an anniversary of the scientist or his achievement. It appears that the international hero will be so well known that it is likely that his image will be shown as a mirror rather than with a meaningful context as a lens. This appears to contradict the trend towards context anticipating a public understanding of science moving to a public awareness of science.

Summary of the study into heroes of science on postage stamps

Each of the countries I have studied has featured heroes of science on their stamps. All have celebrated their own scientists and to a lesser extent foreign scientists who have contributed to the country or the world as a whole. The celebration of a local hero proclaims a positive message about that country and the opportunities and environment for science to succeed; the message is about the country that has raised the hero. The hero is a source of national pride to be honoured. The postal authorities of all countries have used actual images of scientists to validate and justify the message on a postage stamp. In some cases, the individual is apparently considered well known and the image might not require a portrait to substantiate the message that embraces an achievement. The hero of science on a stamp substantiates Scott's observation about "images which provide the possibility of a degree of independent or national assertion" (Scott, 1995. p. 94).

Half of the science stamps I have examined have a direct connection to a known and recognisable scientist. The other half illustrate what I have called science in abstract. Actual space research, as it evolved from the 1950s, is science. There is a less distinct correlation between the visible participants of space research, the cosmonauts, as they were known to the Russian effort, and the astronauts, as they were described by the United States, as scientists. In this study I have recognised these people as mainstream scientists, and this has biased the number of scientists whose achievements have been honoured through the issue of a celebratory postage stamp. It has been Russia and the Eastern Bloc counties, East Germany and Poland where this bias is the most apparent. Interestingly, this bias has resulted in more than the usual number of stamps being issued. Russia has also used the international space programme, in which it has cooperated with the US through NASA, to celebrate US astronauts. This took place before the end of the Cold War, suggesting that Russia was positioning itself to play a part in an emerging global economy.

All countries have celebrated the achievements of foreign scientists. There has been an overlap where scientists have achieved their recorded success in another country. There is a celebrated nucleus of scientists whose discoveries and inventions have changed the world. These international heroes have been acknowledged, largely without an apparent political motive, within the message on postage stamps, as contributors to society.

The representation of scientists and their achievements on stamps

The representation of scientists on postage stamps is changing over time. The change is, to an extent, enabled by changes in the technology of image production and printing. Intaglio methods have been overtaken in many instances by photogravure printing, which enables the use of photographic images. Photographs allow for the image of the scientist to be shown within his or her working environment without requiring further enhancement. Coincident with technology changes there has been a trend toward showing more context associated with scientists' achievements. I have been able to illustrate this trend by examining the images on a number of the heroes of science examples. The trend has been from showing the named portrait of the celebrant, adding context as a small part of the overall image, to a situation where the celebrant's achievement is afforded equal space to a portrait. The cycle can be seen to have travelled a full circle to the situation where the achievement is the image with only a textual reference to the scientist. Somewhere along the way, the development process of a stamp issue has changed. Between the postal authority who initially conceptualise the issue, the stamp selection committee, the designers (several of whom may be asked to submit their unique designs), the printers, and the authority management who make the final decision on the message and how it is presented, something has changed over time. It seems that the understanding of science by the people in the process has changed and this is shown through the awareness of science inherent in the modern design.

The early scientific stamp acknowledged and represented science through the portrait of an individual scientist. I presume that in many countries the scientist would be thought to be known by name or image and achievement so that no elaboration would be required. With an increasing impact of science and technology upon the public, some examinations of the benefits make social, and possibly political statements applicable. It has been the increasing understanding of science that had driven how the message is told. Not all of science is readily comprehensible, however, and many modern developments will be as a result of multiple participants, without a logical figurehead, which allow for representation of the result as the message.

Figure 6.37 shows the relationships, over time, of the models of science communication and the use of context in the representation of science on postage stamps. The stamp designer is saying that we reflect on our stamp images what it is that the general public knows or is thought to know. The timeline shown reflects two significant science communication dates: the Bodmer report in 1985, and the House of Lords report in 2000. The 1990s is the period where I perceive an increase in the context shown in science stamps.

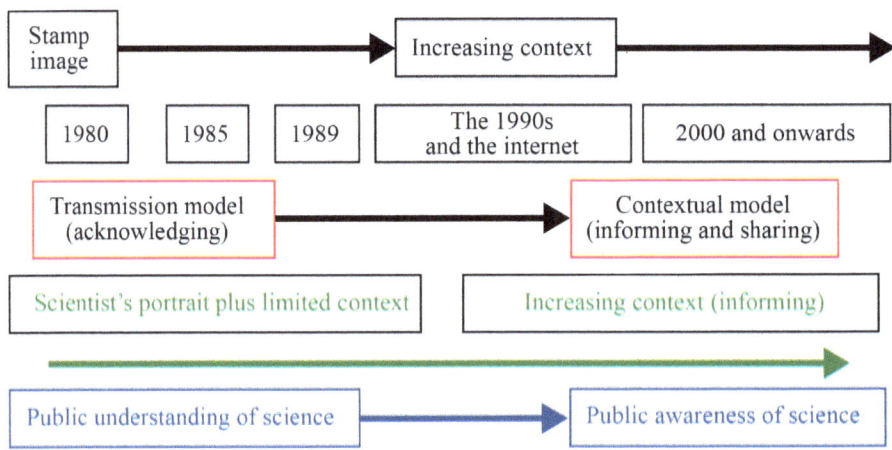

Figure 6.37: The increasing use of context and science communication on postage stamps.
Source: Author's research.

Stamps that contribute to the public awareness of science

A prevalent term in modern literature is the "personification" of science. It is often to be found in the same sentence as "engagement" with science. I am reminded of the remark made by Micheal Zsolt of Australia Post, in which he envisaged Australia Post to have a potential three-second window of engagement with which to grab the attention of someone handling an envelope carrying an Australia Post stamp. Stamps are printed in the millions and there is the chance that the three second window be repeated and reinforced a number of times over the life of that stamp. Everyday exposure to something non-confronting, such as an image of science, may reduce any fear factor that the viewer may have had of that science.

As discussed previously, at the same time as mail deliveries are declining, the number of stamps being issued is increasing. The number of science stamps is also increasing. It is possible that a person now receiving an envelope carrying a stamp will allow it further attention than Michael Zsolt's three seconds. If the design incorporates context as a part of the message, it may appear the more interesting and engage the viewer.

The stamps issued since the mid-1990s have shown a trend towards more context in support of the messages intended for the general public. My affiliation with the Australian National Centre for the Public Awareness of Science has tempted me to seek science stamps conveying a message with a scientific content. Very few actually have a message directed towards public awareness. The examples reproduced in Figures 6.38 to 6.42, from my perspective, address the issue of the public awareness of science. Three of the sets use compilations of the technology of the time to celebrate these as achievements. A sustainability theme, promoting awareness, is also seen in the Great Britain millennium issues, as will be discussed in the Chapter Seven. In 1961, Russia issued a political stamp, shown in Figure 6.38, that might have been titled *Public awareness of science*, although its published title is *Communist Labour Teams – "Adult education"*. Generic figures are shown engaging with a geometrical figure drawn on a blackboard. The problem has their undivided attention. This was one image of three labour teams in their workplace. The iconic hammer and sickle overlaied in red proclaims a government initiative benefitting workers. Its message is more subtle than the overtly political exhortations to achieve production targets that have been examined in Chapter Five. It is also gender balanced.

Figure 6.38: Russia, 1961. ***Communist Labour Teams – "Adult education"***. **Gibbons catalogue # 2633.**
Source: Author's collection.

It took until 2007 for New Zealand to issue a set of stamps that promotes a public awareness of science. The *Clever Kiwis* set, shown in Figure 6.39, shows five quite different aspects of technology, with a New Zealand bias tied by design and the title, shown on each stamp. Also spelled out in each instance is the science and technology achievement, complemented by an engineering drawing and a small image of the object being celebrated. The set is consistent with the New Zealand Post objective of having a national focus and celebrating science and ingenuity as contributors to the social community. It is a set that challenges the viewer to be aware. The objects and inventions chosen to honour *Clever Kiwis* are: the inventor of the electric fence, Bill Gallagher, (1930s); spreadable butter, the brain-child of Robert Norris and David Illingworth of the New Zealand Dairy Research Institute; the mountain buggy, introduced in 1992; the jet boat; William Hamilton (1899–1978), inventor, who was featured as a hero in Figure 6.50; and the animal tranquilliser gun, an invention of Colin Murdoch (1929–2008).

Figure 6.39: New Zealand, 2007. *Clever Kiwis*. Campbell Paterson catalogue # S1051–1055.

Source: Author's collection.

China has facilitated the understanding and awareness of Archimedes' Principle in two sets of two stamps in 2008 and 2010, suggesting the postal authority might be developing a series showing myths of history that apply today. To date, I have not had access to the China 2012 issue programme to confirm it as a series. The first issue (Figure 6.40) is titled *Cao Chong weighs the elephant*. Cao Chong (196–208) lived during the Han Dynasty, was a child prodigy and the son of warlord Cao Cao. The two stamps describe the process: "marking the water level on the boat loaded with the elephant", and "replacing the elephant with weighable objects". The two-part set is directly teaching a science concept. I can see it being used in a classroom to communicate to children who will be engaged by the friendliness of the characters and the directness of the story.

Figure 6.40: China, 2008. *Cao Chong weighs the elephant*. WNS catalogue # CN037–038.08.

Source: Author's collection.

The 2010 title (Figure 6.41) is *Wen Yanbo gets the ball from the hole in the tree*. The historical perspective is that of Wen Yanbo (575–637), a key advisor for Emperors Gaozu and Taizong, and who served as a chancellor during the reign of Emperor Taizong's reign during the Tang Dynasty. Wen Yanbo floats the ball within the hollow tree to the opening into which it was placed.

Figure 6.41: China, 2010. *Wen Yanbo gets the ball from the hole in the tree*. WNS catalogue # CN037-038.10.

Source: Author's collection.

The stamp reproduced as Figure 6.42 has been acknowledged as a contributor to the public awareness of science. This stamp, which was declared the best stamp of 1981 by the Universal Postal Union, dramatically shows the effects of pollution upon a butterfly, a stylised cyprinid fish and a plant and compares these effects with how the living things should look. It incorporates a classic before-and-after symbolism and is highly effective.

Figure 6.42: West Germany, 1981. *Preservation of the environment*. **Gibbons catalogue # 1951.**
Source: Author's collection.

A different public engagement with postage stamps

Royal Mail has sought to engage the public through an app provided to the users of smartphones. The smartphone user photographs an appropriate stamp image, which is recognised by the app, which provides additional data about the stamp to the user. If the stamp image, for example, shows a band the app could respond by playing appropriate music. This approach is being trialed. At this point in time, it is a gimmick for postage stamps, offered sparingly. It is a new method of potential engagement with stamps, not necessarily of science, but it does offer the possibility to move from a one-way communication model to a reactive model where follow-up material can be suited to meet the enquirer's needs.

The results of this chapter combine to help answer questions that are further explored in the next chapter, which examines stamps issued at the time of the millennium and stamps drawing attention to the changing climate.

7. Two Time Capsules

Sharing knowledge powers innovation. To fully realise the social, economic and environmental benefits of our considerable investment in science and research, we must communicate and engage the wider community in science (Department of Industry, 2010).

A postage stamp is a time capsule, a representation of an ideal or a situation at a certain point in time. The point in time is more likely to be the date of decision as to what event, celebration or message is to be circulated than the date of issue because of the lead time for the development of the stamp issue. Nonetheless and regardless of how the representation ages or deteriorates, it is a visual memory and a marker of an idea to be celebrated at a particular time. In this chapter, I examine the stamps issued to celebrate the millennium and those that represent messages about the changing climate.

The stamp issues of 1995–2001: Celebrations of the millennium

In this study, I have been looking at the question: What does the representation of science and scientists on postage stamps convey about the political and cultural necessities of a country at the time of issue? All of my analysis has an implied time criteria, that of the date of issue of the message. This case study affords a conveniently explicit and fairly recent point at which to put down a time marker. Approaching the year 2000, the world had the opportunity for some retrospection and introspection. The millennium was celebrated by some postal authorities to reflect how the country saw its place in the world through the past endeavours of its scientists, their enterprise and effect upon the world. To a much lesser extent, future ambitions were examined through these stamp issues.

Seven of the ten countries selected for this study issued stamps for the millennium. None described the issue that seemed to most concern the world as the year 2000 loomed, the Y2K bug. Scant attention was paid on stamps to the changing climate at this time either.

Australia

Australia Post celebrated 2000 with two stamp issues: a stamp containing a 1999/2000 hologram and a miniature sheet of 25 *Faces of Australia*. The celebration has been an acknowledgement of the date, enhanced technically through the use of the hologram (Figure 7.1). I would not describe the design as inspirational, but it is a colourful representation containing flora as the symbols of the country. The 25 portraits are a graphic illustration of the multicultural nature of the country as it entered the year 2000. The stamps are mirrors, providing no prompt for further thought or action, expect perhaps to admire the diversity of the ethnicity portrayed across all ages. None of the portraits alludes to a scientific theme in any way, although three of the portraits show a uniform.

Figure 7.1: Australia, 1999. *Millennium stamp*. Rennicks catalogue # 1920.
Source: Author's collection.

New Zealand

New Zealand Post embarked upon its millennium celebrations in 1997 with the first issue of a series with the title of *Discoverers,* which featured six of the discoverers of the country. New Zealand Post included the Māori explorers Kupe (c.10th century), describing him as a legendary explorer, and Maui, another historically remembered figure of the Pacific, as well as four European explorers. The discoverers from Europe are Captain James Cook, Frenchmen de Surville, and d'Urville and Abel Tasman of the Dutch East India Company.

The stamps are reproduced in Figure 7.2, shown in date order rather than conventional value order. It is an interesting set, for although tied together by basic format, each explorer is named, the New Zealand symbol for its millennium issues is a constant and all have a sea-blue colour in the image. However, the story is told somewhat differently on each stamp and only four show a representation of the celebrant. Kupe, of legend, is shown carrying the paddle of the two-hulled craft on which he left Hawaii. The background shows a collapsed volcano shell, typical of a New Zealand landscape. Maui is not represented in the image. Set against an aerial view of Cape Reinga are Māori artifacts, the carved manaia and the hei matau traditional fish hook. The *discovery* of New Zealand in 1642 by Abel Tasman is rich in symbolism telling its message of the time. Tasman's likeness and the globe are a fragment of a portrait of Abel Tasman attributed to Jacob Cuyp. A telescope is shown as is part of a drawing "Murderers' Bay", by Isaack Gilsemans, that illustrates when Tasman tried to land at a Māori agricultural settlement and encountering resistance (Flude, 2001). The two stamps showing the date of 1769 highlight the fact that both the British and French were at the time seeking an expansion of land possessions to ferment additional trade opportunities. Both Jean de Surville and James Cook explored the coast of New Zealand that year. The de Surville stamp is illustrated with a map of the top of the North Island charting the route of de Surville's ship, the *St Jean Baptiste*, which included sighting the coastline of New Zealand, and passing James Cook's *Endeavour*, with neither ship sighting the other due to the bad weather. A wooden ship's anchor and a New Zealand headland complete the picture. Over one third of the available surface of the Cook images is from the classic portrait by Nathaniel Dance-Holland, and is complemented by a sextant of the time overlaying a map of the North Island. Dumont d'Urville's image is shown centrally within the image. His ship, *L'Astrolabe*, is shown under sail and rounding a headland, completing the symbolism is a crab. The only reference that might explain the crab is the fact that D'Urville collected biological samples on his voyages. D'Urville is also celebrated on the stamps of the French Antarctic Territory. The variety of formats and extensive use of indigenous symbols move the stamps from being simple mirrors into the lens category, for the interested viewer. The explorers featured cover the full millennium, from the years 1000 to 2000. The indigenous Māori are not again featured in this series of stamps.

Figure 7.2: New Zealand, 1997. *Discoverers*. Campbell Paterson Catalogue # SH79–SH84.
Source: Authors collection.

In subsequent years the series of stamps to celebrate the millennium was continued, although the events celebrated are those of the past 100 years: *1998, A New Beginning*, featuring past waves of immigrants, *1998, Urban Transformation*, showing then and now photographs across the twentieth century, *1999, Nostalgia*, a representation of the icons of New Zealand, *1999, Leading the way*, significant events with New Zealand participation and *2000, First to see the new dawn*. The stamps use both portrait and landscape layouts to tell their message and are consistent in size, with each set of six having a dominant colour. The style incorporates painted and sketched images and, where appropriate, photographs to tell more modern stories. The messages are all mirrors of New Zealand's perceived realities, including the repetition of the claim that Richard Pearse may have made the world's first powered flight. Those realities might, however, be lenses to doubters.

The four scientific achievements depicted on the 1999 set of six stamps, Figure 7.3, and within the later miniature sheet, demonstrate where New Zealand believed it was leading the way are: *Powered flight (1903)*, which shows Richard William Pearse (1877–1953) and the claim that he made at least two flights in 1903 before the Wright Brothers in the United States; *Splitting the atom (1919)* features Ernest, Baron Rutherford of Nelson (1871–1937), the father of nuclear physics; William Hamilton (1899–1978), is recognised as the inventor of the jet boat in 1953; and Sir Edmund Hillary (1919–2008), who with Sherpa Tensing conquered Mount Everest in 1953 and led the New Zealand Antarctic Expedition, 1956–1958, is honoured. The remainder of the set, although not reproduced in the figure, are *Women's suffrage, 1893*, and the political implications of a *Nuclear Free New Zealand 1987* (Di Somma, 1999, pp. 64–69).

New Zealand Post has used two techniques to show its science leading the way. The physical invention is shown for powered flight and the jet boat, but it is the celebrant scientist who dominates the two other science and technology images. The indices are consistent: the name of the country, the stamp value and the title of the issue. Each stamp also carries a symbol, the map of New Zealand with rays emanating from its position on the globe, with this motif carried forward to its 2000 issue, *First to see the new dawn*. The designer has incorporated many images to convey the full message, although the celebrant scientist is not named on any of the stamps. We can only presume that New Zealanders know the stories of Richard Pearse, Lord Rutherford, Bill Hamilton and Sir Edmund Hillary. Both of the artifact images show the engineering drawings behind the inventions, quite literally. The atomic symbol, the New Zealand flag, and the words "splitting the atom" provide a context for the Rutherford celebration. The New Zealand flag is also indicated on the Hillary stamp, with an image of Mount Everest and headlines from 1953 newspapers. The other two stamps, women's suffrage and the declaration of New Zealand as a Nuclear Free Zone, complete the statement that the country claims to have attained the pinnacle of human endeavour for its citizens in the instances documented. The combined effect is mirror although the challenge is issued on each stamp "New Zealand leading the way". Surely, therefore, all are lenses.

Figure 7.3: New Zealand, 1999. The four science stamps from the *Millennium Series V: Leading the way*. Campbell Paterson catalogue # SH104–107.

Source: Author's collection.

Great Britain

Royal Mail undertook a very different approach to its stamp design for the new century and millennium. Two sets of stamps were issued with the generic title of *Millennium series*. Each month in 1999, a set of four, square postage stamps were issued, described as the 'tale' of a featured group of people, a profession or position in society. Some categories were obvious and of interest to this study. The following year, the sets of stamps showed a particular theme. Initially the public proclaimed a dislike for the approach, thinking the stamps dull and uninspired with obtuse images, but time has gilded their reputation for being innovative and embracing the country at the time, as well as being simply different.

Three of the tales are relevant to this case study. They are discussed below with detail provided by the Royal Mail Millennnium Stamps 1999 and 2000 Yearbooks (Davies, 1999, pp. 4–9, 16–21 and 48–53; Davies, 2000, pp. 26–30, 58–61 and 62–67).

Figure 7.4 reproduces the stamps from *The inventors' tale*. The images celebrate: *The Greenwich Meridian*, including John Harrison (1693–1776), the inventor of the first reliable marine chronometer; *Steam power* is celebrated along with James Watt (1736–1819), the image described as robust, muscular and uncompromising; *Photography* is headlined through Henry Fox-Talbot (1800–1841), who initiated the calotype process, with a photographic image of three leaves backlit with rich reds and purples; *Artificial intelligence* celebrates

Alan Mathison Turing (1912–1954) through a sketch of a human head containing ideas. Turing is well known, (so well known that his image is not needed to honour his achievements), and represents computer science. These stamps are all lenses. None of the messages are simple, and all require an explanation to decipher what is being said, for although the title of each stamp is shown (in small text), it is the name of the stamp designer that is given, and not the name of the celebrant.

Figure 7.4: Great Britain, 1999. *The inventors' tale*. Gibbons catalogue # 2069–2072.

Source: Author's collection.

Figure 7.5 shows the stamps of *The patients' tale*. The medical innovations and celebrants are: *Vaccination*, Edward Jenner (1749–1823), "the father of immunology". The Jenner illustration is most arresting as the image of the physician vaccinating a child is shown as a pattern upon the side of the cow, an optical illusion that is not immediately apparent. Included in the image is a small phial suggesting the source of the vaccine. Once seen, it becomes an outstanding example of a lens. The image of *Modern nursing care* celebrates the impact of the nursing practices of Florence Nightingale (1820–1910). A penicillin mould illustrates the experiments of Sir Alexander Fleming (1881–1955). We know it is his work that is being acclaimed as the title of the image is given as *Fleming's penicillin*. Similarly the title *Test-tube baby* explains the message to be one of in vitro fertilisation, examining the research of gynaecologist Patrick Steptoe (1913–1988) and physiologist Robert Edwards (born 1925). In common with all the tales being told, the stamps are all lenses.

Figure 7.5: Great Britain, 1999. *The patients' tale*. Gibbons catalogue # 2080–2083.

Source: Author's collection.

Figure 7.6 reproduces the stamps of *The scientists' tale*. Artist Mark Curtis explores the divide between science and art to illustrate *Decoding DNA*, celebrating the achievement of James Watson (b 1928) and Francis Crick (1916–2004). The imagery investigating *Darwin's theory of evolution*, counterpoints a limestone fossil of Archaeopteryx, the first known bird, with one of Darwin's finches. A photograph of a firework sparkler on a turntable has been used to make the appropriate image for the *Rotation of polarised light by magnetism*, honouring Michael Faraday (1791–1867). The Newton stamp, *Development of astronomical telescope,* features an image of Saturn taken using the Hubble Telescope. The designer wanted to make the point that "all aspects of astronomy inextricably relate to Newton" (Davies, 1999, p. 52).

Figure 7.6: Great Britain, 1999. *The scientists' tale.* Gibbons catalogue # 2102-2105.

Source: Author's collection.

Science pervades every aspect of life, and there is an orphan stamp within *The travellers' tale* set issued in February 1999. It shows Captain James Cook and a New Zealand Māori and is described as illustrating Captain James Cook's voyages. The stamp is reproduced in Figure 6.32.

During the months of 2000, twelve numbered themes were represented with a set of four *Millenium Project* stamps. None of the themes were specically scientific in emphasis, but scientific images were used within three of the themes. The April 2000 *Life and Earth Project* has an image representing solar sensors. The September 2000 *Mind and Matter Project* features an a x-ray of a hand holding a computer mouse. And the October 2000 *Body and Bone Project* features a magnified hen's egg. (Davies, 2000, pp. 26, 58, 62). These stamps are shown in Figure 7.7.

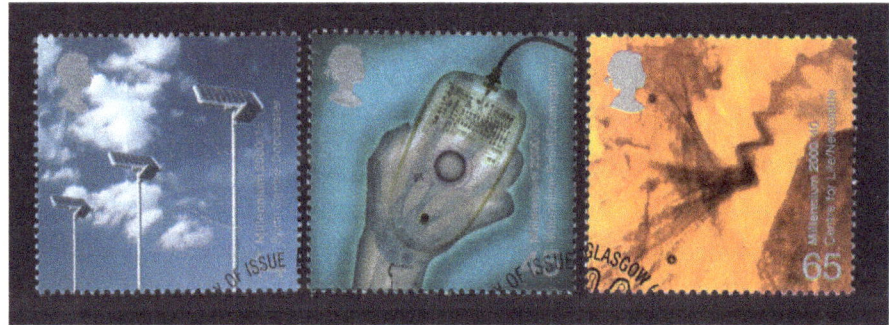

Figure 7.7: Great Britain, 2000. Three stamps from *Millennium Projects*. Gibbons Catalogue # 1453, 2164 and 2169.
Source: Author's collection.

Criticism of the millennium issues of 1999 and 2000 was widespread and filled the letters to the editor sections of a wide range of British publications. None the less, the issues fared well in the Royal Mail its annual favourite stamps poll, condutcted through its monthly *Philatelic Bulletin*. The poll invitation is made in the December issue, with results published the following April. The published results show that the 1999 issue 64p stamp from *The scientists' tale* celebrating Isaac Newton and the Hubble telescope was the favourite. Second favourite scientist was the 26p Charles Darwin stamps showing a modern bird standing over a fossil reptile. The miniature sheet described as *Millennium timekeeper* received no votes. Possibly the symbolism was lost on the general viewing public. *The scientist's tale* was voted the Bulletin readers favourite set of stamps, with *The settlers' tale* coming second. *The inventors' tale* and *The patients' tale* did not poll favourably, receiving one-eighth the votes of the winner (Philatelic Bulletin, 2000, p. 237). In results from the 2000 poll, the *Millennium projects* elicited fewer voters than in the 1999 poll, and the three science-based images shown in Figure 7.6 made no significant impact, with the *Tree and leaf* and *Spirit and faith* sets taking the first two places (Philatelic Bulletin, 2001, p. 244).

Having looked at the millennium issues and some of the reactions to it, it is appropriate to review what Royal Mail set out to achieve with these stamp issues. In the introduction to its annual yearbook, *Royal Mail Millennium Stamps, 1999*, it is stated that:

> ... to mark the Millennium, Royal Mail issued its most extensive and ambitious stamp programme by far, a unique collection of 48 stamps by 48 diverse image-makers, who visually interpreted the past 1,000 years of British history (Davies, 1999, p. 28).

The rationale given for the selection of 12 representative tales was the idea of putting people first in order to make the shared past emotive and relevant. As Jeremy Black, Professor of History at Exeter University, writes:

> There was method in the Millennium madness. This is an exciting account of our shared past. The experience of our history, an attempt to move beyond political history, a public project worthy of the Millennium. Let stamps show you history. After all, more people in Britain saw these stamps in 1999 than read history books (Davies, 1999, p. 28).

The designs of the 12 tales are also discussed in the yearbook (Davies, 1999, pp. 54–55)., which states that the Stamp Advisory Committee (SAC) was asked to consider the concept of using 48 different artists to interpret the issue. Committee member Alan Livingstone notes that "stamps are an important indicator of national attitudes and national aspirations and the SAC is a vital testing ground" (Davies, 1999, p. 55).

One year later, the 2000 annual yearbook explained the choice of subjects and designs for the issues of 48 of the total of 96 stamps that comprise the millennium issue. The introduction to the second book is titled "How to mark the Millennium (part two)" (Davies, 2000).

> The stamps in 2000 celebrate both the Millennium and the massive efforts that have taken place to create four dozen exceptional projects. Together, the two different but complementary sets of 48 stamps constitute Royal Mail's own monument to the millennium. (Davies, 2000, p. 2)

One might surmise this self-enthusiasm to be a response to the public's general dislike of the project. The rhetoric continues:

> It's the vision, rather than the actuality, of the multifarious projects which inspired Royal Mail's approach. Many were still at planning stages when the stamp programme kicked in, many weren't physical monuments so much as intangible ideas aimed at bringing communities together. An inventive, symbolic take on the spirit of the chosen projects proved to be the most viable and creating way forward. (Davies, 2000, p. 4)

Photographs were chosen for the project images on the basis "photography is immediate, vibrant and accessible" (Davies, 2000, p.7).

The millenium stamps also won a design award, with the collection granted "Millennium Product" status by the Design Council, awarded to British products and companies which show "imagination, ingenuity and inspiration" as well as "innovation, creativity and design". Post Office Group Managing Director, Stuart Sweetman, said:

> The stamps are The Post Office's celebration of the Millennium and this award pays tribute to all the superb artists and image-makers who have contributed to this success. Britain, where the postage stamp was invented, has traditionally led the world in stamp design and has broken new ground with this collection.

France

France determined that the new millennium started on 1 January 2001 and looked forward to the event from April 2000 with the issue of the first of four miniature sheets. Like other issuing authorities, France followed a thematic approach (Yvert et Tellier, 2010). Each of these miniature sheets has additional photographic material in the background illustrating the themes. What is interesting about the French millennium issues is that the featured images of science and technology are impersonal, although two iconic space research images are shown.

Sporting Achievements, the first series of the *Twentieth century* issue, was followed in September 2000 by the second series, *Society*. French Society in 2000 is represented by images of the *Declaration of Human Rights, The washing machine, First man on the moon, Women's suffrage* and *Paid holidays*. The third series, *Forms of communication,* was represented by *Television, Public relations, The portable telephone, Radio* and *Multimedia – the compact disc*. The forth series, *Science,* is represented by images of *Penicillin, DNA, Lasers, The first man in space* and *Electronic banking* (Yvert et Tellier, 2010, p. 393).

Figure 7.8 shows the five images used by France to represent science in 2001. The first four subjects are shared with other country's issues, but France is the only authority to highlight *Electronic banking*. The representation of science is quite complex. Three medical breakthroughs from the twentieth century are featured. At the left, a generic figure, (who looks a bit like Ernst Chain, who co-developed the *Penicillin* manufacturing process with Howard Florey), pointing to the bloom in Fleming's petrie dish. A generic figure is embraced by the *DNA* helix. The next two stamps use photographs to tell their message, the first showing the use of a *Laser* for eye surgery, a specific application. Three other countries have shown lasers as a scientific achievement in general terms. Yuri Gagarin is not identified as *The first man in space*, we are expected to know who he is. A module for reentry into the earth's atmosphere module is also shown. The final image of a hand holding an ATM card takes us into modern *Electronic banking*. The stamps are colourful and carry two service fees, the incoming Euro currency and the French currency value it replaces. All stamps are for local usage priced at the base rate for in country service. Two sets of the five stamps were contained within a miniature sheet entitled "Sciences". These include a computer mouse, children lining up for a vaccination, a moon rocket and images of space, a medical imaging perspective of a human head, and the most unusual photograph of Einstein with his tongue out.

Figure 7.8: France, 2001. *The Twentieth Century: Science*, one set of five from the miniature sheet. Gibbons catalogue # MS3756.
Source: Author's collection.

Germany and Poland did not issue any special stamps to celebrate the millennium.

Ireland

The Republic of Ireland, Éire, celebrated with six themed sets of six stamps with the title of *Millenium issue*, issued from the last day of 1999 to the first day of 2001 (Hamilton-Bowen, 2009).

There were no scientists included in *Issue 1, illustrated famous people* or. *Issue II, historic events. Issue III, discoveries* recognised Reverend Nicholas Callan (1799–1864), an Irish electrical scientist and inventor of the induction coil, Albert Einstein (1879–1955), and Galileo (1564–1642). Also within the set, as it was was issued as a miniature sheet containing two sets of featured discoverers, are the Birr Telescope, linked with the work of William Parsons and Thomas Grubb (1800–1878), Marie Curie (1867–1934), and Thomas Alva Edison (1847–1931) performing an early demonstration of the incandescent light bulb.

Iconic photographs have been used to illustrate Einstein and Marie Curie. A reconstruction and photographic approach has also been taken for the four other images and all include a millennium symbol with the figure "2000". All the stamps have been issued at the rate for internal mail within the country. The reconstructions suggest they are mirrors, stating the realities of four significant scientists whilst introducing a lens for the two less-known, (even in Ireland), contributions to science of the Reverend Callan and the Bir telescope. The latter two images shown in Figure 7.9 are quite complex.

Figure 7.9: Ireland, 2000. Three images from *Millennium Issue III, discoveries*. Hibernian catalogue # C985, C980 and C981.

Source: Author's collection.

Issue IV concentrated upon *The arts*, *Issue V* focused on *World Events*, included the Industrial Revolution and modern communications with a science connotation. The final set, *Issue VI, Journeys*, is self-explanatory. The featured journeymen are Marco Polo, Captain James Cook, Burke and Wills, Ernest Shackleton, Charles Lindberg, and space exploration.

With its own delightful sense of self, the Irish Post Office (An Phoist) issued a sheetlet of 15 stamps picturing the *Hurling Team of the Millennium,* a game unique to the country.

Russia

From 1992, the Russian Federation, as might be expected, was a major user of the post to disseminate messages, issuing special stamps to celebrate the turn of the century. The five stamp issues were more conventional and direct than those from Great Britain and the United States. The Russian Federation celebrated the new millennium at the start of year 2001, rather than 2000. The listings below describing five issues are taken from the Stanley Gibbons Stamp Catalogue (Stanley Gibbons Publications, 2010, Part 10).

Achievements of the twentieth century, a set of six stamps that featured aviation, computers, genetics, nuclear energy, space exploration (including a portrait of Yuri Gagarin), and television, was issued in November 1998. The next millennium celebration was issued in March 2000, *The Twentieth Century (first series), Sport*, a set of 12 stamps.

The June 2000 *The Twentieth Century (second series), Science*, was comprised of a set of 12 stamps issued as a miniature sheet with a border containing scientific symbols, and also as separate stamps. Nine images featuring a named scientist and his/her main accomplishment, with three remaining stamps featuring more generic achievements. The range and diversity of the scientists and their scientific achievements is large:

- *Observation of ferromagnetic resonance (1913)*, V. K. Arkadjev (1884–1953), physicist.
- *Theory on plant divergence*, Nikolai Vavilov (1887–1943), botanist and geneticist.
- *Moscow Mathematical School,* Nikolai Luzin (1883–1950), specialist in descriptive set theory.
- *Phenoms theory (1929)* Igor Tamm (1895–1971), physicist and 1958 Nobel laureate.
- *Discovery of liquid helium superfluidity,* Pyotr Kapitsa (1894–1984), 1978 Nobel laureate.
- *Chemical chain reaction theory (1934)*, N. N. Semyonov (1896–1986), the 1956 Nobel laureate.
- *Charged particles in accelerators (1944–1945)*, Vladimer I. Veksler, nuclear physicist.
- *Deciphering of Mayan language (1930s)*, Yuriy Knorozov (1922–1999), ethnographer.
- *Discovery of pogonophora (1975–1977)*, Alekseevich Ivanow (1886–1970).
- *First photograph of the dark side of the moon (1959)*, obtained via Luna 3.
- *Development of quantum electronics (1960s)*.
- *Ethnoliguistic dictionary,* (1955), credited to N. J. Tolstoi.

Figure 7.10 shows four stamps from this issue, *Moscow Mathematical School, Phenoms Theory (1929), Discovery of pogonophora (1975–1977)*, and *First photograph of the dark side of the moon (1959)*. Russia has celebrated only its own scientists in compiling this list. Nine of the 12 images include a sketch of the scientist, given equal prominence as symbols describing his achievement (only males are named). A millennium symbol is shown on each stamp, as are the names and birth dates of the featured scientists, and a short description. The issue includes three stamps of each of four values, showing that there is the intention that the set be used for all classes of service, this is a definitive set with a long shelf-life in the post office. The example shown that does not include a portrait illustrates the mechanism for the "first photograph of the dark side of the moon", clearly raising awareness for science. The stamps praise Russian scientists' ingenuity and are mirrors of the celebrity and lens for the context and a claim for Russian scientific excellence across the broad range of specialities.

Figure 7.10: Russia, 2000. Four stamps from the *Twentieth century (second series), Science*. Gibbons catalogue # 6928–6929 and 6934 and 6935.
Source: Author's collection.

The Twentieth Century (third series), Culture was another set of 12 stamps. The next set, number four in the series, featured *Technology*. This was another set of 12 stamps, issued as a miniature sheet with a border containing appropriate symbols of technology. Similar to the science series, the millennium symbol is shown on each stamp, as is a short description of the technology. The issue includes three stamps of each of four values, again showing that there is the intention that the set be used for all classes of service, another definitive set with a long shelf-life in the post office. The theme and images represent:

- *Medicine*, doctors operating and medical equipment.
- *Construction*, represented by a city skyline.
- *Transport*, with an image of a bus, car and truck.
- *Engineering*, showing a dam, electric pylon and generator.
- *Communication*, telephones, television, rocket and satellite.
- *Space technology*, illustrated by space stations and rocket.
- *Aviation*, with images of civil and military airplanes.
- *Rail transport*, showing steam, diesel and electric trains.
- *Sea transport*, with images of a container ship, sailing ship and a cruise liner.
- *Metallurgy*, represented by a furnace.
- *Oil refining industry*, defined by an oil refinery and truck.

- *Mineral extraction,* with a montage of a truck, conveyor and drill.

Similar to the science series, the technology theme embraces a diverse range of topics and is a reflection of changes over the immediate past century. Older technologies have been ignored.

Figure 7.11 shows four examples of the issue — *Medicine, Engineering, Communication,* and *Space technology* — that show the impact of technology, as a mirror of the effect of more recent technology on everyday life.

Figure 7.11: Russia, 2000. four stamps from *The Twentieth Century (4th series) Technology.* Gibbons catalogue # 6964 and 6967–6969.
Source: Author's collection.

China

Of all the countries looked at for this study, China issued the subtlest set of "wishful thinking" images for its acknowledgement of the western world's *New Millennium* (Scott Publishing Company, 2009). It is worth noting that China continues to use its lunisolar calendar to determine significant event dates. The Kuomintang reconstituted the Republic of China on 10 October 1928, and the Gregorian calendar was officially adopted on 1 January 1929. The People's Republic of China has continued to use the Gregorian calendar since 1949. The millennium message is here developed through combination of five stamps including the following images:

- The sun, moon, date, time and futuristic buildings.
- The earth and a dove (symbol of peace).
- A map, leaf and an infant's hands.
- An electronic circuit board, a human head and the earth.
- The moon, stars and a sun dial.

Figure 7.12 illustrates the only technically-oriented image. The background is a printed circuit board. The main picture is a profile of a human head, and a brain is shown that merges into a representation of the earth highlighted by a star image, perhaps representing artificial intelligence. Two horizontal bands of colour (laser beams, perhaps) cross the profile at the level of the mouth and the throat. This is definitely a hard-edged lens, requiring the viewer to think about the message.

Figure 7.12: China, 2001. Single image from the *New millennium* issue. Gibbons catalogue # 3081.
Source: Author's collection.

United States of America

The United States Postal Services issued 150 special stamps within 10 miniature sheets to celebrate the millennium. Each miniature sheet has a short description of the decade it represents and a background image reflecting the decade. The 150 stamps each have a textual explanation on the back of the stamp of the particular celebration. The images are a mixture of styles, more diverse than was used by Great Britain. My understanding is that this diversity represents a postmodern approach and indeed it has been described as such by critics. If I am to use the expression I need a definition. I offer the following:

> One compact definition is that postmodernism eradicates the boundaries between high and low forms of art, and disrupts the genre's conventions with collision, collage, and fragmentation. Postmodern art holds that all stances are unstable and insincere, and therefore irony, parody, and humor are the only positions that cannot be overturned by critique or revision (Wood, Cole and Gealt, 1989, p. 323).

The significant achievements of science and technology start with the Wright Brothers first controlled air flight, with a stamp and a photograph of the event providing the background of the miniature sheet *The 1900s — The dawn of the twentieth century*. As might be expected, the events celebrated included more science examples as the century progresses. The four achievements of the 1990s are reproduced in Figure 7.13 as examples of the postmodern approach to design (Haskins, 2003) used to celebrate the millennium. The four stamps show to *Computer arts and graphics, Return to space, The World Wide Web* and *Cellular phones*. It is not until one turns *The return to space* stamp over to read on the back that one realises that it celebrates John Glenn's return to space at age 77 on board the shuttle Discovery. The *Cellular phones* stamp shows an African-

American male in a business suit, suggesting that racial inequality in the United States has been overcome by the 1990s. The messages from the 150 stamps are mostly mirrors, with an acknowledgment of scientific advances in each decade.

Figure 7.13: United States, 2000. Four images from *The 1990s — Cold War ends, economy booms*. Each of these stamps has a textual description on is back, gummed side. Scott catalogue # MS3191.
Source: Author's collection.

Evaluating the *Celebrate the century* issues, Associate Professor E. V. Haskins of Boston College presented a paper and later published an article looking at the process behind the United States Postal Service millennium issues. As we have seen, the United States Postal Service issued a total of 150 stamps over a two-year period within ten miniature sheets, each containing 15 images reflecting each decade of the twentieth century. The images for the first five decades were chosen by members of the Citizens' Stamp Advisory Committee appointed by the Postmaster-General. Members of the general public interested enough to pay the postage in order to vote in an open ballot chose the images to represent the second half of the century. A great deal of effort and money was expended to stimulate and sustain public interest in the project, including post offices being turned into mini-museums where mechanical devices counting down to the year 2000 were installed. This was a strong programme of public engagement, including with science, generated through the issue of postage stamps. "*Celebrate the century* was as much about remembering and celebrating as it was about reminding adults and children about what to remember" (Haskins, 2003, p. 1).

Consistency in the messages told through millennium stamps

There is a degree of consistency in the subject matter chosen for the millennium celebrations by the six of the ten countries analysed in this study. Perhaps the most surprising thing is that it has been the twentieth century that has been the focus of the messages. Great Britain and Ireland recognised Captain James Cook and his rediscovery of New Zealand in looking to earlier achievements. Ireland featured the industrial revolution as a world event, along with the Internet, in addition to the explorations of Marco Polo in Asia and the Irish-born Burke crossing of Australia. Germany, Poland did not produce specific issues. China's issue has had an abstract approach. Australia chose to celebrate the diversity of its people. A table listing the general themes contained in the issued millennium stamps is shown in Table 7.1.

Science and technology were strong themes followed by Russia and New Zealand, with attention paid to local celebrities. Great Britain and France looked at social issues, Ireland looked back beyond the twentieth century to remind us of the contribution of Galileo to world knowledge and the United States celebrated, somewhat belatedly, their space successes.

Table 7.1: Themes contained within the six millennium issues.

Theme	Country					
	NZ	GB	France	Eire	Russia	US
Science/named scientists	✓	✓	✓	✓	✓	✓
Medical/health issues		✓	✓		✓	✓
Medical — DNA		✓	✓	✓	✓	
Inventors		✓			✓	✓
Explorers	✓	✓		✓		
Space exploration			✓	✓	✓	✓
Communications/WWW		✓	✓	✓	✓	✓
Computers		✓	✓	✓	✓	✓
Lasers			✓			✓
Environment		✓				
Non-science themes:						
World politics	✓					✓
Womens' vote	✓		✓	✓		
Entertainment		✓	✓			✓
Millennium symbol	✓			✓	✓	

Source: Author's research.

Science and scientists are well represented, with all countries identifying science as significant in their review of the twentieth century. France does not identify any particular scientist, and who can explain the image of Albert Einstein sticking his tongue out in the background of the *Twentieth century – third series* miniature sheet? Russia's set entitled *Science* celebrates both the person and the achievement.

Simple millennium symbols have been adopted by three countries on theirs stamp issued to celebrate the occasion. Russia had two such symbols, shown in Figure 7.14.

Country	New Zealand	Ireland	Russia
Symbol			

Figure 7.14: Millennium symbols used as part of this particular issue of stamps.
Source: Author's collection.

World politics are largely ignored in these issues. New Zealand proclaimed its Nuclear Free stance, and the United States issued 15 stamps under each of the general headings of *The 1940s — World War II transforms America* and *The 1990s — Cold War ends, economy booms*. Evidence, I believe, that postal authorities have not used the millennium celebration to send messages of bad news.

Where a country has produced a series of issues to celebrate the change of millennium, they have generally reviewed the twentieth century and its achievements and discoveries. It is an historical theme that has been adopted by six of the ten countries of my study. These countries have chosen to feature science achievements, which provide a framework for my analysis. The science featured has emphasised current pursuits, implying a look to the future, although the United States did divide the tzwentieth century into a review by decade. Scientists of international fame are featured but the Russian Federation, for one, has managed to find local personalities to illustrate their twentieth-century achievements. It has been a surprise that the changing climate and environmental sustainability do not feature, except in one case, discussed below. Culture seems to have been ignored; perhaps it was too hard to visualise? Great Britain, France and the United States adopted a postmodern approach to design and mixed representation to suit the subject matter.

Only Great Britain seems to have set out to highlight the changing climate and the need for sustainability. The *Project* stamps of the year 2000 "celebrated both the millennium and the massive efforts that have taken place to create four dozen exceptional projects" (Davies, 2000, p. 2). The overall categories of the projects "educate; connect communities or ideas; renew (landscape or spirit); or sustain (environment or culture)" (Davies, 2000, p. 7). I have made the judgement that four of the set of 48 stamps directly reflect a changing climate concern (see Table 7.1) against the theme of environment. A few stamps have been issued prior to

2000 that are relevant to the changing climate. These are shown in Figure 7.18 and are described in the associated narrative. The changing climate story is taken up in the next section of this chapter.

Could the British 2000 millennium projects provide a marker for a different approach to the scientific messages conveyed on British postage stamps? It does seem appropriate to ask the question, as the science messages on stamps today increasingly show the context of achievement in addition to showing the celebrant, although the transition from portrait to context has been earlier in other countries.

The use of creative, innovative designs, including postmodern elements and sophisticated production techniques, also suggest that the millennium is being celebrated as a point in time in a modern society looking to the future, rather than as a historical reflection of the past 1,000 years. I do not believe the millennium issues provide a strong evidence of political and cultural awareness. Perhaps the postal administrations, in trying to involve the general public in determining the messages to be told at the time, have lost perspective. The science that is featured is the science of the time with little regard to the future.

Messages about the changing climate

> Scientists have warned about the "greenhouse effect" for years. Now it is no longer a scientific nightmare. It has arrived. (Sydney Morning Herald, 1988, cited in Taylor, 2012)

Climate Change is one of the issues of the day. The issues of the changing climate are current today and, in reality, have been so for 30 years. One might expect that the message of the changing climate, sustainability and environmental conservation would be a constant theme on postage stamps. It is not. Berglez, as detailed in in this section, contends that the changing climate is being recognised as an international problem rather than as a localised situation that any one country can solve. This contention will be tested by looking at the stamps of the countries in this study.

Reviewing the data, I am struck by the various titles given to the messages through stamps that draw attention to the environmental situation, such as conservation, pollution and protection of animal and plant species. This section concentrates on the messages that express concern about the changing climate and environmental protection. My hypothesis is that conjecture and challenge on the stamps shown within this section will be lenses. The postal authority has issued these stamps to change public perception and behaviour.

Taylor has studied the subject of framing within a science communication context and has used the *Sydney Morning Herald* citation above as an introduction to her description of her PhD thesis (Taylor, 2012). Her focus is on public knowledge, the ensuing public dialogue, and government action with regard to climate change from the late-1980s to 2001. She has shown that the actual science available, published by the Intergovernmental Panel on Climate Change (IPCC), since 1988 has been consistent, but that the public rhetoric of the Australian Government has varied with the politics of the day. It has been political necessity that has driven (or not driven) government action with regard to the changing climate.

In the late 1980s, Australian Prime Minister Bob Hawke called for action on global warming, as reflected in the Australia Post stamp issue shown in Figure 7.15. The four images and text cover the spectrum of the ongoing concern: C*onserve our soil*, *Precious pure air*, W*ater is precious* and *Conserve energy*. The stamp design and format are consistent and form a cohesive whole, although the pre-payment of mail service is from local to international mail. These stamps are strong lenses to promote public changes of behaviour prompted by sophisticated images evoking curiosity in understanding the messages.

Figure 7.15: Australia, 1985. *Conservation of soil, air, water and energy.* **Renniks catalogue # 887–900.**
Source: Author's collection.

In keeping with Taylor's conclusions about a long gap, between the 1980s and 2000s, in public concern regarding climate change, it would be 19 years before Australia Post again reviewed the changing climate situation, highlighting the importance of renewable energy. The communication from two other Prime Ministers, Paul Keating and John Howard, changed dramatically "from expressing good understanding and a will to take action, to a confused and conflicted debate with clear correlations to the national response" (Taylor, 2012). The four stamps shown in Figure 7.16 illustrate four possible sources of renewable energy but with less of a call for action than the earlier set. The *Renewable energies*: *solar, wind, hydro* and *biomass* stamps are named in the foreground but the theme is shown in a very small font. The images are mirrors of the technologies, with an element of lens, although the environmental message is diluted. The emphasis for the action for change has been lost because of the change in political motivation.

Figure 7.16: Australia, 2004. *Renewable energy: solar, wind, hydro and biomass.* Renniks catalogue # 2333–2337.

Source: Author's collection.

After Howard's 11-year term in office, in 2007, Kevin Rudd's Labor Government was elected to power. Rudd had previously declared that the resolution of climate change was the "world's greatest moral challenge", and one of his first actions was to ratify, on behalf of Australia, the Kyoto Protocol which was adopted by the United Nations ten years in 1997. Australia Post responded to that challenge with the stamp set entitled *Living green* which was also issued as a prestige stamp booklet with background information text. During 2007–2008, Australia was in drought control mode and the first stamp, *Save water,* shown in Figure 7.17, was particularly pertinent. All the stamps in the issue were for local service, with the three other prompts being *Reduce waste*, *Travel smart* and (again) *Save energy*. As Ericken *et al*. assert: "Climate change and its alleviation is the underlying subject of this issue" (Ericksen *et al.*, 2008, p. 32). 2008 was also important in Australian terms as the Garnaut Climate Change Review was published, stressing the urgency and expense of the primary challenge "to end the linkage between economic growth and emissions of greenhouse gases" (Ericksen *et al.*, 2008, p. 32). The images are strong and simple lenses sending a very clear message.

Figure 7.17: Australia, 2008. *Living green: save water, reduce waste, travel smart and save energy*. **WNS** catalogue # AU055–058.08.

Source: Author's collection.

The fourth set in the series (Figure 7.18) celebrates Earth Hour, an Australian initiative in 2007 and now a worldwide event organised by the World Wildlife Fund (WWF). Earth Hour is held annually on the last Saturday of March encouraging households and businesses to turn off non-essential lights for an hour to raise awareness about the need to take action on the changing climate. Australia Post has chosen three animals to represent the fact that all life on earth is threatened by a change in the climate, with design driven by the concept of protection. The leadbeaters possum, shown on the green 55 cent image, is an endangered species in its natural habitat in Victoria. The orangutan, whose image is on the international stamp, is in danger due to the deforestation of Indonesia and Borneo. The third image carries the main semiotic message in that in the context of the three stamps the "owl represents the wisdom of taking timely action against global warming" (Ericksen *et al.*, 2009). This fourth set is the first from Australia Post that embraces Berglez's concept of space in taking a worldview to tell its Australian message (Berglez, 2012). These Australian stamps confirm Taylor's basic thesis that changes in political motivation determine priorities in the narrative supporting, or denying, the need for public policy.

Figure 7.18: Australia, 2009. *Earth Hour*. WNS Catalogue # AU019–021.09.
Source: Author's collection.

New Zealand has yet to issue a stamp with a changing climate message. The most obvious reason for this is the New Zealand Post policy of promoting tourism. It would not be appropriate to associate New Zealand with problems of sustainability through its stamp issues.

Having looked at two countries, one that has featured stamps on the subject of climate change, and one that has not, I now wish to discuss two papers in the public arena to help with understanding the changing climate story.

Berglez, of Örebro University, Sweden is investigating climate change, questioning the ways that politics and political action are represented in media and citizen discourses. One media characteristic he reviews is the multi-level appearances of spaces, political actors, powers and identities in news texts. By "space" he refers to the multi-faceted geography in which changing climate stories tend to look at concerns on a worldwide basis, simultaneously occurring in separate places worldwide (Berglez, 2012).

Schroeder, Boykoff and Spiers (2012) have studied representations in climate negotiations, stating:

> Over the past five decades, multilateral institutions and global governance mechanisms have emerged to address these environmental challenges but with mixed success. To avert irreversible global change, fundamental and radical transformations of existing governance practices are now needed. Indeed, state function has shifted from "a role based in constitutional powers toward a role of coordination and fusion of public and private resources" where states have become "increasingly dependent on other social actors" (Schroeder Boykoff and Spiers, 2012).

In the next section I build upon the Australian experience to trace how the postal authority has represented the changing climate and environmental protection through its stamp issues and show that changing climate messages on stamps tend to be universal and not specific to a particular country.

Great Britain

Figure 7.19 shows a set of four stamps issued by Royal Mail in 1992. Children's competition paintings, sponsored through a well-known BBC Childrens' Television programme, have been used as the images with the subject representation explained in text. The four issues raised are A*cid rain, Ozone layer, Greenhouse effect* and *Bird of hope*. The Berglez international dimension, space, is apparent. A review of the issue includes the observations that:

> Green issues ignored geographical and national boundaries ... Children were as aware of — and concerned about — environmental problems affecting far-away countries as they were about issues close to home. The message to their elders was unmistakable — do something about it now, because by the time we are old enough it may be too late (Shackleton, 1992, p. 30).

Figure 7.19: Great Britain, 1992. *Protection of the environment.*Gibbons catalogue # 1629–1632.
Source: Author's collection.

In 2000, Royal Mail issued 48 stamps, of which one billion were printed, with the theme of "Projects", which covered the UK at the time of the Millennium.

> Projects were ordered into four broad categories according to whether they primarily: educate; connect communities or ideas; renew, (landscape or spirit); or sustain (environment and culture) (Davies, 2000, p. 4).

The Great Britain millennium issues, two years each of 48 stamps embracing a wide spectrum of issues, is examined separately above. Not since 1992, (Figure 7.18), had Great Britain sent so many themed messages that embraced such a wide spectrum of issues, some of which directly and indirectly send a message about protection of the environment. Figure 7.20 shows the complete set of 48 stamps for 2000, four stamps in each theme were issued every month of the year.

Figure 7.20: Great Britain, 2000. The complete set of *Millennium Project*. Gibbons catalogue # 2125–2132, 2134–2145, 2148–2159 and 2162–2177.
Source: Author's collection.

France

France recognisied the need for conservation of natural resources quite early, using a worldwide perspective with a stamp issued in 1978. But, similar to other countries, France had a 21-year gap before returning to the subject. The 2005 *Environmental Charter,* an initiative of President Jacques Chirac is celebrated on the second image in Figure 7.21. The charter explains:

> In their battles against climate change, genetically modified organisms, and nuclear reprocessing, the French now enjoy the support of an "environment charter" amended to the country's constitution. "Decisions made responding to today's needs should not compromise the capacity of future generations and other populations to satisfy their own needs," the document's preamble proclaims. In 10 articles, it then outlines a series of environmental rights and responsibilities incumbent on the French people, ranging from the right to access information about the environment to an obligation upon political leaders to promote sustainable development (grist.org, 2005).

The 2010 issue incorporates the image of Marianne, the iconic mother-figure of France, supporting the need for *Water protection*.

Figure 7.21: France, 1978, 2005 and 2010. *Energy conservation*; *Environmental charter* and *Water protection.* Gibbons catalogue # 2269, 4110 and WNS # FR021.10.

Source: Author's collection.

With an initial set issued in 2008 entitled *Sustainable ideas,* the French postal administration has embraced the preservation theme. Examples of three of the set of ten stamps are shown in Figure 7.22. Not all the images are science stamps by my definition, but they do show thoroughness in pursing the concept. A world map is included with the admonition that the whole world is a site worth preserving, although the text suggests the world should be preserved for tourism. The stamps have been produced as self-adhesive items. This set shows the service level to be provided, priority letter up to 20 grams, as France Poste follows other authorities in defining the service, for which Poste take a responsibility in perpetuity, at today's price.

Figure 7.22: France, 2008. *Sustainable ideas*. WNS # FR089-098.08.
Source: Author's collection.

The 2010 issue, *Water protection: Environment and conservation* is a long set of 12 stamps, again in the self-adhesive format and lettre prioritaire, three of which are shown in Figure 7.23. The design for each image has a three-dimensional icon emerging from a two-dimensional surface, emphasising a thought-provoking strong impression. All are excellent examples of a lens seeking to influence water conservation.

Figure 7.23: France, 2010. *Water protection: Environment and conservation*. WNS catalogue # FR022-033.10.
Source: Author's collection.

Germany

West Germany, has been the most consistent country in producing stamps with the message of conservation, and have issued the earliest message on the theme (Schroeder, Boykoff and Spiers, 2012). The images in Figure 7.24 show the German Environmental Conference emblem and highlight concerns with *Waste, Water, Noise* and *Air*.

7. Two Time Capsules

Figure 7.24: West Germany, 1973. *Protection of the environment.*
Gibbons catalogue # 1666–1669.
Source: Author's collection.

During the 1980s, West Germany issued, almost annually, a stamp related to a conservation theme. The United Nations' Intergovernmental Panel on Climate Change issued its first report in 1990. In 1995 West Germany proclaimed the *First Conference of Signatories to the General Convention on Climate, Berlin*, as shown in Figure 7.25. The other images contained within Figure 7.25 are examples of strongly contrasted styles from the period of West Germany and reunified Germany from 1991.

Figure 7.25: West Germany 1980, Germany 1996 and 1995.
Nature conservation; Environmental protection: preservation of tropical habitats; First Conference of Signatories to the General Convention on Climate, Berlin. Gibbons catalogue # 1951, 2626 and 2729.
Source: Author's collection.

Figure 7.26 contains an image that links environmental protection with renewable energy.

Figure 7.26: Germany, 2004. *Environmental protection and renewable energy.* Gibbons catalogue # 3252.

Source: Author's collection.

Reunified Germany issued another distinctive set in 2004, shown in Figure 7.27, which features five climate zones. What makes this set distinctive is the fact that the complete set of images is included on each stamp, the main climate zone being considered and the four others in a much smaller scale on the side that incorporates the service price. The stamps also incorporate a premium upon the postal rate. The locations (climatic zones) are worldwide, emphasising Berglez's notion of universal imperatives for the environment. The locations are: Greenland (arctic), Tibet (mountain), Mecklenburg-vorpommern in the German Baltic (temperate), Sahara (Desert) and Galapagos Islands (tropics).

Figure 7.27: Germany, 2004. *The environment: 5 climate zones.* Gibbons catalogue # 3293–3297.

Source: Author's collection.

Ireland

The Irish stamp issues of the 1970s were highly stylised. The 1970 issue shown in Figure 7.28 uniquely celebrates *European Conservation Year*. No other European authority has chosen to represent this early acknowledgement of the need for conservation.

Figure 7.28: Ireland, 1970 and 1979. *European Conservation Year* and *Energy conservation*. Hibernian catalogue # C143–144 and C282.

Source: Author's collection.

Again we witness, through the Irish Post Office, the acknowledgement of a problem brought to the public attention that is then ignored for 32 years. Renewable energy is the theme of the 2011 set shown in Figure 7.29. The images feature the technologies of solar, wind, wave, hydro and biomass on self-adhesive stamps.

Figure 7.29: Ireland, 2011. *Renewable energy technologies*. Hibernian Catalogue # not yet assigned.

Source: Author's collection.

Poland

Poland has issued four stamps with general messages of conservation and the changing climate since 2001, three of which are shown in Figure 7.30. The Polish 2001 *Europa, water resources* image imaginatively addresses a worldwide problem. The United Nations Conference stamp again incorporates the earth to illustrate the changing climate is a world issue.

Figure 7.30: Poland, 2001, 2007 and 2008. *Europa, water resources*; *Earth Day*; and *United Nations Conference on Climate Change, Poznan*. **Gibbons catalogue # 3910, 4265 and WNS catalogue # PL062.08.**
Source: Author's collection.

Russia

The Russian Federation has published stamps to send messages and a concern in the changing climate since 1989. The first set of three (Figure 7.31) is titled *Nature conservation,* with world maps indicating areas of concern for each of the environments that constitute the images shown. The environments are *Forests, Arctic preservation* and *Anti-desertification*. I have classified this set as being appropriate to an analysis looking at the changing climate as the maps define the areas of concern on a worldwide basis. The se-tenant maps and text mean that these are focused lenses in intention. A worldwide concern is shown confirming Berglez's contention that this is a feature of the representation of the changing climate message.

Figure 7.31: Russia, 1989. *Nature conservation*, maps indicating areas of concern. Gibbons catalogue # 5967–5969.

Source: Author's collection.

The next issue (Figure 7.32), from 1991, sends messages related to *Environmental protection* by presenting specific areas of Russian environmental concern and naming an animal threatened by the changing climate: the Bell Tower in Volga and the sturgeon; Lake Baikal and the sable; and the dried bed of the Aral Sea and the saiga. The saiga is a critically endangered antelope which originally inhabited a vast area of the Eurasian steppe zone. The initial reaction to the two outside images, a view of a water-enveloped city and ships stranded in a desert environment because of desertification, is confronting. These are strong lenses focused, for the first time, on a local basis.

Figure 7.32: Russia, 1991. *Environmental protection*. Gibbons catalogue # 6227–6229.

Source: Author's collection.

Russia celebrated World Ozone Layer Day, and the 10th anniversary of Montreal Protocol on the reduction of use of chlorofluorcarbons with a single stamp in 1997 (Figure 7.33). The image shows radiation, represented by straight lines hitting the earth. This day was proclaimed in 1994 by the United Nations General Assembly and it is celebrated annually. The image here is a dramatic lens.

Figure 7.33: Russia, 1997. *World Ozone Layer Day.* Gibbons catalogue # 6711.

Source: Author's collection.

The Russian Federation postal authority once again features a worldwide map to mark the World Climate Change Conference held in Moscow in 2003 (Figure 7.34).

Figure 7.34: Russia, 2003. *World Climate Change Conference, Moscow.* Gibbons catalogue # 7204.

Source: Author's collection.

The 2005 set shown in Figure 7.35 invoked the description of earth as seen by the Russians from space as "the light-blue planet" and features water, describing the theme of water preservation through five dramatic photographs without narrative text.

Figure 7.35: Russia, 2005. *The Earth, light-blue planet.* **WNS catalogue # MS7369.**
Source: Author's collection.

China

As discussed in previous chapters, China issued stamps that looked outwards from the 1980s, but its first stamp projecting an environment message was issued in 1992 (Figure 7.36), celebrating the 20th anniversary of a United Nations' initiative. China is a signatory to the Kyoto Protocol, but as a non-Annex I country is not required to limit greenhouse gas emissions. The People's Republic of China is an active participant in climate change talks and other multilateral environmental negotiations, and claims to take environmental challenges seriously, although it is pushing for the developed world to help developing countries to a greater extent, as it argued at the 2009 Copenhagen Conference of the United Nations Framework Convention on Climate Change.

The English name of the issuing country is shown on stamps from 1992 as China frees itself into a global economy.

Figure 7.36: China, 1992. *World Environment Day, 20th anniversary of United Nations Environment Conference.* **Gibbons catalogue # 3796.**

Source: Author's collection.

One would have to think that China is committed to protecting the environment, with its issue of a set in two parts during 2002–2004, which includes one really interesting message, that of maintaining a low birth weight. I cannot think of another country that would promote this as an ideal. Shown in Figure 7.37, series one features a global perspective on *Maintaining low birth weight*, *Forest production*, *Mineral resources protection*, *Air pollution prevention*, *Water resources protection*, and *Ocean protection*. Series two features *desertification control and prevention,* and *biodiversity protection*.

Figure 7.37: China, 2002 and 2004. *Environmental Protection* **(series 1) and** *Environmental Protection* **(series 2). Gibbons catalogue # 4664–4674.**

Source: Author's collection.

China celebrated World Earth Day with a uniquely round stamp with four star-shaped perforations in the perforated circle (Figure 7.38). The main image is the globe being cared for by multicolored hands. Also included within this figure is a 2010 issue featuring two ambitions: *Low carbon development* and *A green future*.

7. Two Time Capsules

Figure 7.38: China, 2005 and 2010. *World Earth Day* and *Energy conservation and emission reduction*. Gibbons catalogue # 4976 and 5459–5460.

Source: Author's collection.

United States

The United States is the only country in the world not to have ratified the 1997 Kyoto Protocol, despite having recognised the need to advise the public of anti-pollution concerns as early as 1970. Since then, there was a gap of 25 years before the changing climate again featured on US postage stamps. Childrens' paintings have been used in 1995 and 2001 issues (Figure 7.39) suggesting, perhaps, that it is the younger generation who will be affected by the impacts of the changing climate and that their involvement is crucial. The images used in the earlier issue are *Earth clean-up*, *Solar energy*, *Tree planting* and *Beach clean-up*. The *final stamp here, Stampin' the future* is from a set of four titled *Loving the world*.

Figure 7.39: United States, 1995 and 2001. *Earth Day* and *Stampin' the future*. Scott catalogue # 2951–2954 and 3415.

Source: Author's collection.

261

The Representation of Science and Scientists on Postage Stamps

Summary of stamps that have included messages representing the changing climate and environmental protection

Figure 7.40 is a bar chart plotted to show the number of stamps that have been issued since 1970 classified as having a changing climate or a sustainability message. As noted earlier, has been West Germany and, more recently, reunified Germany that has kept the message consistent, with a specific issue in 13 of the 21 years between 1980–2000 and 24% of all stamps with these messages from the ten counties studied. As might be anticipated, the number of changing climate and environmental protection stamps increases significantly after the millennium.

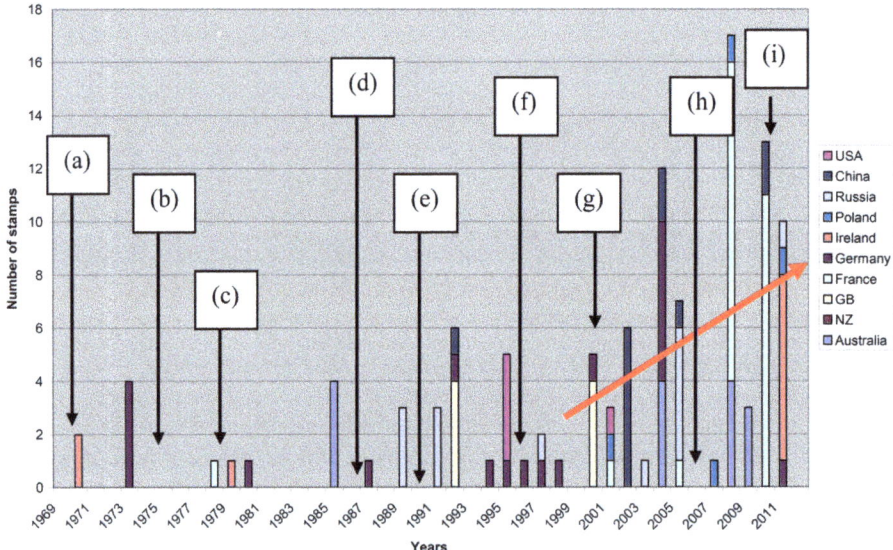

Figure 7.40: The number of stamps issued over the past 50 years with the subject of the changing climate and environmental protection.

a) 1970: First world environment Day.
b) 1975: Discovery of damage to the ozone layer.
c) 1979: World Climate Change Research Programme launched.
d) 1987: The UN Montreal protocol on "Substances that deplete the Ozone Layer".
e) 1990: 1st IPCC report.
f) 1997: Kyoto Protocol signed by 37 countries, including the European Union.
g) 2001: US President Bush denounces Kyoto.
h) 2007: 4th IPCC report. Ex-Vice President Gore and IPCC win Nobel Peace Prize. Australia signs the Kyoto Protocol. The European Union confirms its environmental policies with the Lisbon Treaty.
i) 2011: United Nations Climate Change Conference, Durban.

The red arrow reflects the increase in stamps that tell the message of the changing climate since 2000. They have been issued, as illustrated above, from Australia (3 sets, 11 stamps), Germany (9 issues, 13 stamps), Ireland 2 issues in 2011 (seven stamps), Russia (3 issues, seven stamps), and China (4 issues, eleven stamps). Source: Author's research.

Postal authorities have publicised the prevailing scientific and public concern over climate change for 50 years and seemingly ignored the issue at times when government imperatives have changed, as is shown by Taylor (2012).

Germany has been the most consistent messenger, publicising the changing environment with issues — mostly single stamps — during 9 years between 1987 and 2011. France has issued the most stamps, with long sets in 2008 and 2010. The long gaps of inactivity between postal administrations first showing its concern and the subsequent issue seems to follow the Australian example highlighted by Taylor in her review of the changing political emphasis on the changing climate over time.

Every stamp has used science in abstract to convey its message. No image is personalised, with the exception of three of the US childrens' paintings that include generic figures. Human hands are shown by three countries, signaling that the solution is in our hands. Photographic images are a common thread in drawing attention to the changing climate.

All the images contrive to send a message to make us aware of the problem. As pointed out in the study of the millennium stamp issues, the changing climate was largely ignored and was not, it appears, a political priority in 2000, despite the unchanging IPCC Reports and the Kyoto Protocol.

The stamps examined within this changing climate and environmental protection section are all strong lenses.

Berglez's contention that the changing climate is seen as a worldwide problem, taking place in multifaceted geography in which interrelated processes and practices occur in separate places worldwide is proven through the use of maps and the earth shown as a globe by almost every issuer. The problems are acknowledged, but not necessarily accepted through policy changes and action by individual countries.

8. Discussion

> The postage stamps devoted to scientific themes play a vital role in promoting positive attitudes toward science on the part of the general public. (De Young, 1986, p. 2)

Previous academic research has been conducted to gauge the communication potential of the postage stamps of a few specific countries and some regional bases. This research concentrated on such issues as geopolitics and the creation of national icons, particularly at a political level. But none had specifically looked at how science messages are communicated. This has been the opportunity afforded by my study. Having defined what I believe constitutes a science message on a stamp I have used a case study approach to answer 'how' and 'why' science and scientists are represented on postage stamps. My focus has been investigating the subject as a historical and contemporary phenomenon, examining the context and meaning of the science messages being communicated to the public. Science has been a message theme of postage stamps for approaching 90 years, and it is being adopted more frequently as postal authorities increase the number of postage stamps being issued. In addition to the increasing number of stamps being issued, I have shown that postal administrations are proportionally increasing the number of science themed stamps in general. As discussed in Chapter Three, more than one in ten of all stamps published during the last five years shows an aspect of science to convey its message, compared to the long-term average of one in 14. Science as a theme registers with the public and is in keeping with the increasing public awareness of science shown through the increasing sophistication of the context to tell the stamps' messages.

As an extension to Scott's (1995) supposition that in the early part of the twentieth century European countries sought to establish national icons through their representation on stamps, Hymans (2004) explored the idea of the creation of national icons in the case of celebrities celebrated on European paper money, finding iconographic similarity across space and iconographic differences across time. Hymans did not look solely at scientists, but his conclusions have a lot in common with this study. He observes that single named celebrants are depicted with a traditional portrait, calling more attention to the named figure than the money itself. On banknotes, unnamed figures were grouped together, a trend shown on stamps that I have described as generic or aspirational figures. As it was banknote printers who were first employed to design and print postage stamps, there will be a common thread for the early issues, before the scientist appeared as the message carrier. Hymans saw a move towards a common design style developing on European paper money. I do not see that trend on postage stamps, which is perhaps not unusual, given the small size of a stamp; or it could be that the postal administrations themselves constantly change design in order to maintain collector loyalty. There is reason to think that the images shown on postage stamps will be expected to be of lasting interest as stamps were collected from the beginning of their existence.

I have discussed at length my use of the mirror and lens classifications as a way of determining the how and why of a stamp issue. The image content determines the classification. Not only has the level and integrity of context within science messages on stamps increased over time, the understanding of the science message shown by the stamp designer has increased over time. This shows that the designer's intent to engage with science has increased along with their awareness.

Over the past few decades there has been a change in the way science is conducted, from an individual to a research team approach (Searle, 2011). This may be a factor in the reduction in the number of individual scientist portraits that are used as the message image, these being replaced by the contextual image which tells the whole story of the team's achievement. But the need for celebrity to personalise messages is clearly still current. This study shows that, except for the large techno-nationalistic grand design projects previously discussed, a lead figure is still heavily featured. Messages of significant achievements, particularly if described as breakthroughs — such as *Medical Breakthroughs* (Figure 8.1) — are acknowledged specifically identifying a named scientist as a textual focus. Each of the stamps in the figure has a dramatic image, (these are definitely lenses), but each also attributes the achievement to a single person. The text is small, but as a lens the viewer might pursue their enquiry by reading the text, albeit with the aid of a microscope.

There is another specific time development that should be considered. Many countries have featured the world wide web as a significant achievement celebrated at the millennium. The internet has become a content-rich source of information. I do not think it a coincidence that context on stamps has become richer since the mid-1990s. It seems reasonable to believe that stamp designers and stamp selection committees have used the internet as a source of data and inspiration.

Frewer, whose work focuses on Japanese stamps, is unequivocal: "postage stamps are a medium of communication" (Frewer, 2002). He does not differentiate between the subjects of messages. My initial thoughts regarding Stocklmayer's (2013) communication model were that the postage stamp would only ever be regarded as a one-way information medium that typically informs the reader, listener or viewer, informs policy, and/or affects attitudes and possibly behaviour. This study has shown that the apparent intent of some postal administrations is to do more than just inform the viewer. Political messages certainly have informed policy. Stamp messages on public health warnings, for example, are intended to affect attitudes and behaviour as well as to build an awareness of science.

8. Discussion

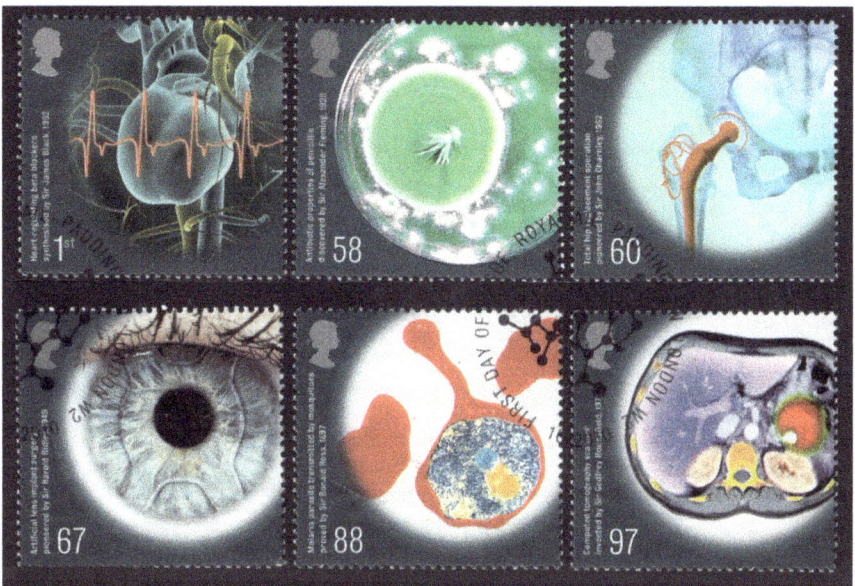

Figure 8.1: Great Britain, 2010. *Medical Breakthroughs.* WNS catalogue # GB130–135.10.

Source: Author's collection.

Nisbet and Scheufele perceive a "paradigm shift within the scientific community that involves a movement away from a singular focus on science literacy as both the culprit and the solution to conflicts over science in society" (Nisbet and Scheufele, 2009). They review how public health messages intertwine with the changing climate messages (although not necessarily on postage stamps), for "going broad" to generate attention and interest among non-elite audiences. As this study shows, the postage stamp sits at the intersection of politics and public values, raising the issues of the role of science and technology in everyday life.

The taxonomy I adopted has helped to illuminate my research. Half of the 4,800 science stamps I looked at had a named and recognised celebrant as the main image, many including a corresponding message stating that scientist's achievement. Stamps identifying the scientist have been issued, usually on the anniversary of the achievement or the scientist's life dates. The other half of the science stamps had representations of science that was not personalised by association with a particular. Such stamps and their messages are also event or anniversary oriented, contain a scientific image, or are institutional; for example, public health campaigns whose message was science in support of public health. Each country has a different profile when it comes to its choice of how to construct a scientific message. Within these two basic classifications, it is possible to identify why the message was sent at that particular time. From my sample of ten countries, Germany and Russia celebrate anniversary dates the most commonly, France and Ireland the least frequently.

Gimmicks have occasionally been used to draw attention to a stamp's message. Unconventional shapes are sometimes used, and I have even seen stamp surfaces impregnated with perfumes to attract attention. Russia has issued three stamps printed on aluminium film to emphasise technical advancements that are a by-product of space research. The development and use of context to put scientific achievements into perspective has been the most influential factor and change noted during this study. The use of photographs and the freeing up of design to dramatise the space race and the historical perspectives of the millennium stamps have been carried through onto today's palette of message-telling options.

The research questions

1. What does the representation of science and scientists on postage stamps convey about the political and cultural imperatives of a country at the time of issue?

As discussed in Chapter Two, several researchers have described how stamps have been designed to establish icons to represent a country, (Altman, 1991; Scott, 1995) with these icons becoming a part of the culture. My study has focused upon particular political messages using science as the vehicle. The ten countries studied all have a different profile when it comes to using science as the message carrier, as was suggested by Petress (1991). The political imperatives have been studied in depth in Chapter Four. What was particularly noticeable was the dearth of science on stamps in Russia following the deaths of Stalin and as a result of the dissolution of the USSR in 1991. China has exhibited a similar pattern of not issuing science stamps during the Cultural Revolution, which is understandable in the light of what was happening, but with a complete change after the death of Mao Zedong. Such circumstances were predicted by Raento (2006). Major changes in approach are not discernable in the more established countries of the west. Cultural aspects, as explored by Kress and Van Leeuwen (2006) and Scott (2005) are, however, apparent when dissecting stylistic and taxonomic differences, country by country. The established European countries were those who, in the eighteenth and nineteenth centuries, sought additional colonies for trading opportunities and dispatched mariners and explorers. Great Britain, France, Germany, Spain, Portugal, Holland, Poland and Russia

8. Discussion

expanded their empires, and countries like Australia, New Zealand and, to a certain extent, the United States were initially explored, subdued and colonised to gratify European ambitions. Many of these historical facts have been recorded on postage stamps reflecting key events in messages designed for civic education and nation building, as suggested by Stoetzer (1953). The seamen and navigators who conducted voyages of discovery were the polymaths of that world and were later followed by explorer-scientists of various disciplines who undertook the exploration of the land masses.

These same European countries and China had a history of scientific study and therefore access to a range of local celebrities whose achievements, if recorded on stamps, fulfill the criteria for recognition at a national level. Reminding the public of scientific achievement fulfills the message roles in a historical context and celebrates those achievements on a local and international basis. Series of stamps honouring specific scientists are issued regularly over time by all postal authorities to confirm the beneficial aspects of science to the public. Lorimer and Scannel define "mass communication as a means of providing information, images ... to large numbers of people ... choosing to attend to an information source" (Lorimer and Scannel, 1994). I would assert that the postage stamp is such a form of mass communications.

The case studies in this study illustrate how specific themes are developed and sustained for a country to advertise its objectives and place in the world. One-party countries have used the postage stamp to directly publicise political issues and ambitions, as typified through such statements as Five Year Plans. Stamps have also been used to promote priorities in scientific achievement, pitting one country against another. The message of science supporting the general public is also publicised through public health issues prevalent at the time of issue.

My taxonomy has shown that half of all science stamps utilise a named and recognised scientist, very often a portrait, to focus their message. The other half use images of science in abstract and may use generic figures to illustrate how the science is being promoted. Generic figures are used when the message is aspirational and looking forward to a science-enriched world. The Royal Society ad hoc Group (1985) recognised that "subliminal sources of scientific information can make a significant contribution to public understanding that extends well beyond overtly scientific items" (Royal Society ad hoc Group, 1985, p. 33). Stamps are in this category.

Within my research I have raised the time of issue as an important factor. Science on stamps has been a reality for 90 years. Science has certainly changed over that period and its representation has had to change and develop. Much of the data I have been examining is historical. The images and context used are pertinent to the time the stamp is issued. That does not mean that they will be readily understood at a later time or by a later generation.

I have conducted two small-scale surveys in relation to determining the applicability of the mirror/lens argument. My evaluation of mirror/lens is biased by my familiarity with the data set and the historical circumstances of the stamp issue. At the time of issue, the context will have been meaningful reflecting what the designer expected to be understood from the message they were asked to convey. These surveys and the wide span of results on a mirror/lens scale recorded by the participants confirmed that how the images are viewed is very subjective, context and time are interrelated. A mirror or lens judgement needs to be made with understanding of the time, space and social context. This is too stringent a requirement to be used as a firm classification method but is nevertheless fruitful in determining designers' intentions to reflect reality or challenge perceptions.

2. Are there constraints and/or conventions imposed on the stamp issuing authorities which predicate their publication of scientific constructs?

There are a few constraints to the subject matter that might be published by postal authorites. We know that the issuing authority is enfranchised, through its policy statement, to send positive messages that reflect well upon the country. Negative messages are not deliberately sent, although many aspects of the impact of climate change have been shown over the past 20 years reflecting the concern the world purported to be showing at the time, before change of leadership and political reality framed new agendas, as recorded by Taylor (2012). Some of the approaches to sustainability have also been featured, such as wind turbines and solar panels. The public face of concern, such as public service sustainability and Earth Hour, has been published, but there is no discernable issuing policy over time, reflecting postal administrations' uncertainty as to what is important at any time.

One constraint upon the subject matter might be the time taken to develop a stamp issue. From my discussions with three postal administrations, I learned that development for an issue is scheduled over a two-year period. The inhouse authority research team will have, possibly with prompts from the general public, determined which events and anniversaries fall two years hence, in order to start a preliminary plan to allow time for institutional consultation as to how the event or anniversary might be celebrated. The 2010 Great Britain *350th anniversary of the Royal Society,* is an example of joint development between the Society and Royal Mail to celebrate ten scientists in a set each representing a significant contribution within a 35-year period. Celebration of significant historical events are criteria defined for the postal administrations as an element of their charters. Rose (1980) emphasised this function as an important aspect of the design process which has continued to the present day.

The main constraint has been the fact that living persons are generally excluded as stamp subjects. What also appears to be a real concern for the postal authority is to have established a person's credibility and place in history before he or she is commemorated. It is obvious that in being able to commemorate a sporting achievement the following day, as has been achieved by a few authorities, that pre-planning of formats and distribution have overcome the stamp development time lag.

Because of the changing technologies that are available to the issuing authorities, I have suggested that opportunities for bespoke stamp issue are a reality. Authorities such as Royal Mail and Australia Post have observed the rule of not celebrating living persons other than the Royal Family. But the world is changing and, according to Jeffries, "in recent years the rule has not so much been bent, but more twisted, snapped and then brushed under the carpet (Jeffries, 2011). I note that Australia Post has issued an annual series of stamps, titled *Australian Legends,* and declared the selection criteria for this award would ensure it be given to "inspirational" living Australians who "exemplify tenacity, imagination, perseverance, devotion, integrity and compassion" (Fahour, 2012, p. 2). Directly applicable to my study are the sets of five stamps, *Australian Legends Award — Medical Science* 2002, and *Australian Legends Award — Medical Specialists* 2012, in which photographic images were taken specifically for Australia Post and complemented with descriptive booklets. The use of scientists' images hardly elevates them to the status of celebrity sports stars, but does place science securely within the real world.

3. Are changing perceptions of public awareness and attitudes to science mirrored on postage stamps?

In general, the evidence provides a positive answer to this question, although it might be a fairly recent phenomenon for some countries. Evaluating the messages contained within the millennium issues, I was struck by the lack of representation of concern for what was becoming a critical political issue for the world, that of the changing climate and environmental protection. I have discussed this at some length in Chapter Seven, but note again here that these issues are mostly portrayed on stamps as a worldwide problem. Countries have not put their hands up to accept responsibility for the changing climate or put forward a local solution on their stamps.

The relationship between science and technology and society has changed over time. With the industrial revolution and the movement of people from a rural to an urban environment based upon manufacture, came the requirement that science be accommodated, learned across multiple disciplines as an everyday part of life. An understanding of science was encouraged through the establishment of scientific institutions that aspired to share their knowledge. Today it is expected that an acceptance of science by the general public is a factor in the development of nation building and improvement of lifestyle. These phases can be traced through the images of science on messages within postage stamps. There has been a change from the use of the portrait of a famous scientist towards a contextual interpretation of scientific achievements, although it appears there is little opportunity for the development of dialogue between the stamp issuing authority and the users of postage stamps. There is little and limited anecdotal evidence that some themes are unpopular on stamps — bad news or an overtly technical approach to telling the message — that might cause less public acceptance and stamps being rejected in the post office itself.

It is true to say, however, that there has been an international move to engage the public with science through any number of mechanisms, which include public controversies over science and technology; science communication in the mass media; science museums, aquaria, planetaria, zoological parks, botanical gardens, fixed and mobile science exhibits; science festivals; science fairs in schools and social groups; science education for adults; consumer education; public tours of research and development parks, manufacturing companies; science in popular culture; and science in textbooks and classrooms. To which, one might add messages on postage stamps.

As far back as 1972, it was remarked that the stamp designer must use images to clarify abstracts, such as the idea of evolution. I believe the increasing trend of adding more meaningful context to stamp images is a reflection of the increasing public awareness of science. Today's public expectation of science, as enunciated by Masters (2012), means that a stamp has to show integrity and validity. I have not been able to put a definitive date on when the more appropriate scientific context appeared more regularly on postage stamps, but this looks to have occurred in the mid-1990s.

Great Britain's Postmaster General Benn in 1965 expressed his objective of having a more liberal subject policy (Chapman, 1994). Parker endorsed this objective in 2011 as an extension of public patronage to the arts. There is some indication of this occuring where the postal administration has used formal portraits of scientists to illustrate their message.

Stamps are generally printed for the most common of the services that a post office will expect to provide, to simplify over the counter sale of the required postage fee. It is logical to expect, therefore, that most viewers will only ever see one stamp on an envelope as it passes through their hands, and upon delivery. As we have seen, some authorities have adopted a policy to convey a particular message through a single stamp, while others use a set of stamps to provide a more complete sequence of images to tell its message. But it has to be assumed that single stamps from the set will be used by the general public, so each stamp in the set also has to tell a meaningful part of the message. There are opportunities to tell a more expansive message when the design can be incorporated into a miniature sheet, which might contain multiple copies of one stamp or different images and also use the selvedge, (any space outside of the stamp perforations), for explanatory image or text. But even then, individual stamps will be torn out of the miniature sheet for everyday use. It will only be the stamp collector who will regularly reconstitute a miniature sheet of the used stamps after they have completed their primary function. The same collector will probably have also sent the full sheet adhered to an envelope to a colleague collector if stamps that have been fiscally used are their objective.

Science and technology on stamps is communicated by designers who are not scientists, although they may be guided by institutions and professional communicators. This may be one reason why mathematical and chemical formulae are not that common. I suspect, however, that this might be because of the low level of visual appeal and a negative attitude towards mathematics. There have certainly been instances when formulae have been challenged after publication. The strong negative reaction to the perceived error on the 2008 Gerty Cory Nobel Prize acknowledgement (Figure 6.63) is evidence that various publics are well aware of the science images on their stamps and the ideal of the accuracy of the science.

Traditional frontiers between communicative contexts in science have been adapted to meet the challenges of the physical constraints of telling a message in a limited space. Stamp designers have been able to use some unexpected technologies in the telling of the message. Holograms have been used to show more than a single image on the stamp's surface, and stamps have been coated with a scent that is released when the surface is scratched. The only instance of which I am aware that the stamps in a set have been cut to simulate a jigsaw is the Great Britain *Darwin birth bicentenary* celebration of 2009, which illustrates the complexity and diversity of evolution. The association of science developments with everyday artifacts is possible because of the public's increasing familiarity with the beneficial byproducts of scientific research.

4. Have stamps been issued that contribute to the public awareness of science?

The increasing use of context has led to science stamps of the lens type that encourage people to look at the message to understand what it is saying, to think, to be curious about science, and to stimulate an engagement with science through the contextual design. Stamps as lenses are used to tell important public messages, although the reception of that message will be an individual response. I have selected a few stamps that I believe have contributed to the public awareness of science. The choice is subjective and seems to follow several formats. One consists of retrospectives with images of several discoveries or inventions that encapsulate the message to be told. Irish Post have, for example, named and shown with context the achievements of four scientists they proclaim as Irish, whereas historically they have been identified as English. The Chinese set of ancient implements shown as the early versions of today's technologies seeks an awareness of science and technology. The second format is different. I have selected two stamp sets that have used a naïve, some might say childish approach to illustrate Archimedes Principle that represent an educational and scientific awareness to solve everyday problems. The Russian stamp selected, uniquely, shows adults enjoying an industrial training session.

Every commemorative science stamp raises a scientific issue. It will be in the eye of the beholder to determine whether it raises an awareness of science

Public health issues have been highlighted on postage stamps as a separate genre. They are science in that they are designed to promote awareness in the expectation that medical science will provide a resolution to the problem. Techniques of medical examination have been explained through the images and text. I have not highlighted issues such as drug abuse, AIDS and SARS, although the images used have been to the point, and certainly raise awareness of a problem, although not necessarily declaring a solution.

Emergent themes

There are a number of additional emergent themes in the representation of science and scientists on postage stamps:

1. Science on postage stamps is used as a communication device to convey messages to the general public, although these messages may be designed to tell a variety of different stories.
2. Stamps convey good news, only very infrequently will bad news be publicised.

8. Discussion

3. Countries show significant differences in their approaches to the representation of science on postage stamps.

4. As governments have looked to the development of science and technology as a legitimate political theme, science has become a subject theme in its own right.

5. The representation of science is subject to framing to meet changing political requirements.

6. The impact of female scientists has been largely ignored on postage stamps, as was pointed out more generally by Davies (2010).

7. Public health issues, although not strictly science, have been publicised on stamps and have been brought to the publics' attention with some innovative, dramatic images. The general theme has been that of science finding a solution to the issue.

8. Postage stamps have been used as a charity collection medium over many years, to further support victims of disasters and public health issues.

9. Postage stamps, and science on stamps, have become available in a variety of guises as the postal authorities seek to optimise revenues from stamps at a time of fewer mail items being carried through the mail, prepaid through the purchase of a stamp.

10. More postage stamps are being issued, year-on-year, including science on stamps.

11. Postal authorities may also use series of issues to enforce the message over time.

12. The integrity of design within a set of stamps appears not to be jeopardised to cater for the international market.

13. Changing technologies may influence design and the messages to be told on future stamps.

14. The identification of the message on a stamp as a mirror or a lens has proved a useful way of looking at stamp images and the designer's expectation to engage with the public.

The Representation of Science and Scientists on Postage Stamps

1. Science on postage stamps is used as a communication device to convey messages to the general public, although these messages may be designed to tell a variety of different stories.

The earliest stamps used a national icon such as the monarch, a coat of arms or a flag, as the image on the label that prepaid the postal fee. In 1888, however, the Colony of New South Wales changed the use of the image on a stamp in order to commemorate a specific event. New South Wales celebrated the centenary of the First Fleet's arrival in Australia and used images that recorded Sydney scenes, local flora and fauna and figures of historical interest within a set of eight values. Captain James Cook was featured on that set as the discoverer of Australia. The message was directed to the western world at the end of the nineteenth century. The message completely ignored the fact that the country had been inhabited by the original Australians for perhaps as long as 100,000 years. History has shown that this message, reinforcing the concept of the ruling elite of the time, causes hostility in some. But in terms of my research, the issue of the stamp had seemed appropriate at the time to the postal administration of the time.

Discussing science on stamps, Ivor Masters of New Zealand Post told me that "the stamp's message must be kept relevant in today's world". New Zealand Post has as an objective that the New Zealand stamp will always convey a message that "has world-wide integrity and validity".

2. Stamps convey good news, only very infrequently will bad news be publicised.

I have discussed this briefly in Chapter Four. The one stamp I could find that told bad news is shown as Figure 4.76, which records *The fifth anniversary of the Chernobyl Power Station disaster* and reflects through its image radioactive particles killing vegetation. Not recording bad news items is one criterion that distinguishes stamps from other media. I have related within my study how the Great British 1998 set of *Endangered species* proved unpopular with the British public because the graphs behind the image showed a decline in species numbers. Subsequent available market research had indicated that the buying public was not keen to buy stamps that conveyed negative news and this appears to have influenced future Royal Mail policy.

The evidence from this study supports Clapper (Barr 1993). The buying public did not expect to be confronted with a negative message and did not like the way the message was framed. Attitude reinforcement is also apparent when considering the frequency with which postal administrations celebrate scientific achievement on a regular basis on anniversary dates.

3. Countries show significant differences in their approaches to the representation of science on postage stamps.

Postal administrations adhere to their stated policies in their issue of postage stamps. They continue to promote science issues that might be described as nation-building, or a form of civic education, building a sense of national identity through recognition of local heroes of science and acknowledging scientific achievements that have increased public wellbeing. The regular publication of science and scientists on stamps contributes to the continuing stories which parallel the country's aspirations.

The composition of the message on the stamp includes not only the interrelationship of semiotic signs but also the expectation that science will be included when appropriate. The four Antarctic authorities, essentially recording the activities of the scant human population of the continent resident for scientific research purposes, record the unique flora and fauna. This somewhat limited palette of visualisation has resulted in an Antarctic style of representation, (see Chapter Four). One might, cynically, argue that the flora and fauna issues are aimed at the thematic collector, but this does not invalidate the intention of reflecting the real world, a mirror of the Antarctic.

4. As governments have looked to the development of science and technology as a legitimate political theme, science has become a subject theme in its own right.

This premise has been argued in Chapter Four. Additionally, most countries have established a regular publishing schedule to recognise prominent citizens. Increasingly, these citizens include scientists.

5. The representation of science is subject to framing to meet changing political requirements.

Public policy and government ambitions directly conveyed to the general public through messages on stamps has been examined. The publishing of targets for Five Year Plans is an obvious example. The framing phenomenon has been discussed in Chapter Seven with particular regard to the changing climate, in which I have been able to show a pattern of messages consistent across countries that illustrate Taylor's (2012) hypothesis. Taylor has shown that the changing climate was a subject being discussed freely in the 1980s, but which lost currency as solutions became mired in politics and became a subject that was ignored for 30 years. The story on stamps has been consistent with Taylor's contention across the countries studied.

6. The impact of female scientists has been largely ignored on postage stamps.

Some early research preceding my definition of a hero of science immediately suggested that some male scientists have a truly international appeal and profile deemed suitable for carrying a science message. Apart from Marie Curie and Florence Nightingale, female scientists have only been recognised on a local stage, and the actual numbers are small. In Chapter Three, I have shown that, to the end of 2011, only 62 female scientists have been shown on postage stamps, compared to some 2,000 named male scientists. Approximately half of the named female scientists have been celebrated on the stamps of Russia and the United States. Generic female workers have been used, principally by East Germany and China, within political messages emphasising the need for females in an increasingly science-dependent work force, but the numbers are still small. East Germany's 14 representations of female generic figures are more that one-third of the total. If there are any themes inherent in the use of female scientists it is in showing their role in medical practice and as Nobel Prize winners.

The participation rate of women as scientists has increased with time, particularly since the middle of the twentieth century. This means that there are not yet many achievements on the calendar of anniversaries for celebration. Another factor might be the fact that there has been a written, (in the case of the United States), or unwritten convention not to celebrate living persons on postage stamps. Russia broke the mould when, as part of its advertising of space successes from 1961, female cosmonauts were given equal exposure to their male colleagues and were celebrated on subsequent anniversaries of the flights. Australia is one other country that has rescinded that convention and has featured female scientists in each of its *Celebrating medical science* sets of stamps since 1995. The ratio of female representation to male has been 2:3 in the Australia Post issues of 2008 and 2012.

The United States has marked the recognition of its three female Nobel Prize winners for science with recent issues in 2005, 2008 and 2011. The earlier stamps have been shown in Chapter Six.

8. Discussion

7. Public health issues, although not strictly science, have been brought to public attention with a range of interpretations to illustrate the message being conveyed on the stamp.

As discussed in the previous section, specific public health issues such as AIDS and the SARS epidemic have been raised by a number of countries. Cancer and the harm caused by smoking have also been brought to the public's attention on stamps. However, no country is going to want to point out deficiencies in its lifestyle to the world in general until a resolution has been found. Stamps are, in a real sense, advertisements promoting the country. Stoetzer described "the stamp as a vivid expression of that country's culture and civilization and of its ideas and ideals" (Stoetzer, 1953, p. 1).

8. Postage stamps have been used as a charity collection medium over many years, more recently to further support victims of natural disasters and public health issues.

New Zealand and France have for years issued stamps with a charity premium included in the service sales price, with these premiums going to specific health initiatives. The United States *Breast cancer research* stamp of 1998, (Figure 4.78) has raised US$76.3 million for research.

9. Postage stamps, and science on stamps, have become available in a variety of guises as the postal authorities seek to optimise revenues, from stamps at a time of fewer mail items being carried through the mail, pre-paid through the purchase of a stamp.

The stamp units of the postal services are expected to be profitable. Richard Breckon, the historian of Australia Post, explained to me that every issue is expected to generate revenue of AU$2 million. The stamps themselves are issued in a variety of formats including: individually over the post office counter; in a stamp booklet of stamps with the same denomination; or as a set in a package in the post office or for sale by direct mail. The stamps may also be available in a presentation pack, expected to be a casual purchase as a gift for someone else, in a prestige stamp booklet. These two last items contain background information about the stamp issue and its design. The postal authority also publishes an annual yearbook with all the stamps that have been issued in that year.

Direct mail purchase is available for all these formats. Each new format allows the designer to move to a more sophisticated level of the message being told and afford, in the case of the yearbook, an occasion to explain the ideas behind their design.

Philatelists are targeted as potential purchasers of each format and Australia Post, for one, has experimented with minor modifications to the standard stamp issue, by printing a few copies of the subject stamp without perforations, for example. In this study I have made no distinction as to the source of the science stamp.

To optimise the revenue for a particular stamp issue, the postal authority may additionally produce articles following the stamp theme, using copies of the images, such as jigsaw puzzles, umbrellas, and drinking mugs. Taking this into account, it is possible that some stamp subjects might be ignored if the subject does not lend itself to an additional revenue opportunity.

10. More postage stamps are being issued, year-on-year, including science on stamps.

The taxonomy results described in Chapter Three show the increase in the numbers of stamps being issued annually compared to the average over many years and the increases in the past ten years. It is expected that this trend will continue. Jones concluded that "more stamps are celebrating popular culture than was the case in the past and less celebrating high culture" (Jones, 2004, p. 80). Humour has been used, but has not necessarily proved to be successful. The increasing use of photographs seems to suggest a realism that keeps pace with the growing awareness of the scientific world to place science in context.

11. Postal authorities may also use series of issues to enforce the message over time.

I am able to discern some pattern in an individual country's issuing policy. All countries publish commemorative stamps to celebrate annual events and activities, such as religious festivals, on an annual basis. In the shorter term, three to five years, for example, a country might explore a topic as far as it remains topical and valid. I am thinking here of Great Britain's *Action for species* series, each of ten stamps that were issued annually between 2007 and 2011. Other series which define a country over time include, for example, the US Postal Service celebrating *Distinguished Americans*, including scientists, for its definitive stamps, and its current series of *American scientists* sets of four stamps. Other administrations follow this lead.

The Russian Post Office, while issuing single stamps for more scientists than any other country, has sustained a perspective upon airplanes through its several issues and series celebrating the design prowess of its aircraft designers in addition to its celebration of space research.

12. The integrity of design within a set of stamps appears not to be jeopardised to cater for the international market.

It might be expected that a country issuing a set of stamps of different values would use images to suit each of the service levels. I have seen no evidence within a set of stamps that a well known image has been selected for the prepayment of international mail instead of a less known image,. However, some single stamps and miniature sheets are priced at a value that will be applicable for international use or the specialist collector.

13. Changing technologies may influence design and the messages to be told on future stamps.

The policies of the individual postal authorities have been shown to be different. Some issue a set of stamps with different images to develop the message being told, rather than telling the message with a single image. Postal authorities may also use series of issues to enforce the message over time. The integrity of design within a set of stamps appears not to be jeopardised to cater for the international market. By this I mean, I can not discern that the higher, international service fee stamps change the image within the set theme to be obviously for overseas consumption. Individual commemorative stamps are different and as one-offs it might be argued that these messages are designed for a foreign audience rather than a local one. Changing technologies may influence design and the messages told on future stamps.

It is evident that digital technologies are being trialed by postal administrations that will reduce the time frame for the development of new stamps which might then keep pace with changing events in order to constitute a living history. Postal administrations are seeking new markets for stamps and living histories might constitute new opportunities. Royal Mail has already implemented augmented reality within a few stamp issues and it is possible that further tests will lead to it being used regularly. This will require integrating message design across a variety of mediums and possibly change the stamp image to become the vehicle to a smart device.

14. The identification of the message on a stamp as a mirror or a lens has proved a useful way of looking at stamp images.

As my own familiarity with the representation of science has developed, I have been able to appreciate the mirror or lens perspective to confirm how and why a stamp image has been designed. That familiarity has given me a historical perspective that is not necessarily available to an inexperienced observer. Two small surveys, in which I asked participants to define a few science stamps as either mirrors or lenses, showed a complete range of answers. The crucial factors were time and the clarity of the image. The message on a stamp is geared to the time and circumstance of its issue. The impact of an image of the first successful moon landing is greater in the immediacy of the event than when it is celebrated 50 years later. The survey suggested a busy design that requires unraveling will be determined to be more of a lens than a mirror.

What of the future representation of science and scientists on postage stamps?

Bespoke stamps, printed at source, have been in development for 20 years to produce receipts for postal service over the counter or, increasingly, through kiosks at or near the post office. At this time, touch-screens are replacing the older vending machines that delivered a set value stamp from a coil of preprinted stamps. The new vending mechanisms allow for the weighing of an article and prepayment for a particular item.

I have mentioned how the privatisation of postal services has introduced the opportunity for new players to experiment with how science is represented. Two other developments are significant, the first being the almost instantaneous printing of stamps, so far only used to celebrate sporting successes. Another option comes from a revenue-making activity of the post office who offer the personalisation of postage stamps, through which anyone can have a personal image printed alongside a valid postage stamp for the dispatch of regular mail items.

The New Zealand postal administration has printed specific company advertisements, called commercial advertising labels (CALs). The CAL is valid for postal use at the same standard service fee as a stamp. There is no reason to suggest that unique science not be publicised, or a particular scientist be used to promote a commercial activity.

As discussed earlier, issuing authorities are operating in an environment where less conventional mail is being sent, although with increasing internet the need for conventional parcel post, and labels, is growing.

I believe the threat to the philatelist market from the counter-printed stamp is the fact that the label can be of any value, so that the collector desires to acquire a complete set will be frustrated by an infinite number of possibilities. Commentators have suggested that the changes taking place may have created a "feeling of anger and cynicism" (Deering, 2011). Deering observes:

> The world is changing and the post office network must evolve with it. I believe there will be stamps for a long time to come, but just different stamps — perhaps mostly produced in a different way for a modern automated and self-service world. (Deering, 2011, p. 39).

There is a clear indication that in the twenty-first century the postage stamp is still seen as a potent marketing tool to extol the virtues of the state, and that science is a denominator in the equation. The number of postage stamps being printed on an annual basis is growing and the number of science stamps being issued is also on the rise.

One other opportunity exists for instantaneous stamp production. In 2000, during the Sydney Olympic Games, Australia Post made available, in most local post offices, a stamp that commemorated the Australian gold medal winner of the day before. The stamps were sold in sheets of 10 stamps, and included a head and shoulder portrait of the celebrant with their gold medal prominent as a part of the image. If it was a team medal, the whole team was shown. Details of the celebrants' events were also included. It was a world first. Australia Post repeated the process for the 2004 Athens Olympic, the 2006 Melbourne Commonwealth Games, the 2008 Beijing Games, and the 2012 London Games.

During the London 2012 Olympic Games, Royal Mail was emulating Australia Post's feat and showing its own "Team GB" winners on specially-issued stamps the day after their gold medal-winning events. With this issue, Royal Mail is moving the technology one step further, featuring, where possible, an action photograph of the gold medal winner performing. Australia Post is continuing its head and shoulders format, which now looks somewhat antiquated against the Royal Mail equivalent. What is important is that the application works, albeit that the background indexation of the stamp has been preprinted. Adding the icon, the image, is achieved digitally and almost instantaneously prior to printing and distribution of the postage stamp. The respected commentator Peter Jennings has written: "The convention of not showing living people, other than members of the Royal Family on stamps has gone forever" (Jennings, 2012). Jennings' forecast has been proved true in the first few months of 2013, when Royal Mail celebrated *50 years of Dr Who* by publishing the faces, but not the names of 11 actors who have played the part, including those still living. The latest 2013 set of *Eminent Britons* has not taken the opportunity to include living persons. However, I can report that Royal Mail has celebrated the 2013 Wimbledon Tennis Champion, Scotsman Andy Murray, with four stamps in a miniature sheet on 8 August 2013.

I forecast that, subject to strict content guidelines and license arrangement, the time will come when commercial enterprises will want to issue their own personalised postage stamp in real-time to celebrate achievements. There is no reason to think that science and technology would not be included.

Privitisation of mail services has allowed non-governmental operators to publish their own stamps (receipts). In 2011, Deutsche Post/DHL issued a set of 12 celebrating German mathematicians of the twentieth century, with a strong focus on the theoretical number development of David Hilbert (1862–1943). One stamp, which I would have liked to include in Chapter Six, shows the dates that a theory of relativity were announced by Einstein and Hilbert, just five days apart. In 2012 CityPost Hannover issued a set celebrating the stepped reckoner of von Leibnitz (1646–1716) that looks similar to a conventional stamp, except that each stamp incorporates a tracking barcode. These NGO issues open a new source of study in the awareness of the history of science.

9. Conclusions

In their "optical consistency" and efficiency in fostering a "national" community stamps are comparable to maps and statistics in that they are both "*fixed* as a representational form and *movable* across territory as inscribed on paper" (Raento, 2006).

Introduction

The postage stamp exists as a viable lightweight and portable mechanism to prepay for postal services and transmit information around the world. In looking at the science on stamps messages of ten different countries, this study shows such examination can provide a truly international perspective. Science messages on stamps fulfill a wide spectrum of objectives, from the celebratory acknowledgement of the achievements of a scientist to the massive technonationalistic developments of the twenty-first century. The basic reason for raising a science issue might be nation-building, civic education, notification of political decisions, celebration of events and anniversaries, public health advice, and even propaganda, for which science is the medium of choice. The science issues raised are far-ranging and serve as a continuing reminder to the wider community by providing perspective and meaning to the role of science and technology.

The prevailing science issues of the twentieth century have been examined from a science communication perspective. I have shown that stamps as a communication medium have developed from providing simple portraits of the scientist or sketches of equipment, to experimental narratives and images providing a picture of what a country believes to be relevant at the time. Flight, communications and the world wide web, computers and lasers, DNA and medical advances, vie with space exploration as the most popular themes, combining iconographic images with minimal textual content. Reading and understanding the correct messages requires knowledge of science at the time of issue. Knowing the date of issue permits a detailed story of the history of science to be developed. Postage stamps are major media artifacts.

Semiotics has proved to be a most useful perspective through which to study the qualitative role of signs in human culture and social interaction as a process to analyse how the message representing science is conveyed to the public. Through the use of signs, ideas, ideals, objects and philosophy are disseminated, mainly in a non-confrontational way, although there have been stamp issues that do confront to challenge and change behaviours.

On a single or multi-country basis, the images and messages on science themed stamps can be traced to illustrate unique developments as they took place. For example, 'Flight' is an illustrated story taking us from the earliest experimental balloons to the space rocket, featured incrementally on the stamps of many countries as the interest in flight spread, especially in the early-twentieth century. Using the image of the airplane, for the new service, 'air mail', was celebrated at the time when it was making news. Illustrating a Cold War political bias, space exploration was shown on a flight-by-flight basis by Russia and Eastern Bloc countries to emphasise the scientific enterprise of the space race. The number of stamps issued by the USSR and its cohorts was inflated by as much as 30–40% during this period. The subject was treated more circumspectly by the United States and the west in general.

The life of the stamp and message is infinite, at least as long as the world's 30 million collectors remain committed to the pastime and are serviced by the philatelic trade who reprint the images on an annual basis.

The representation of science and scientists on postage stamps

Postal administrations have, on average, chosen a science issue for inclusion in one stamp in ten. The proportion of science issues to all others has been on the increase from around the year 2000. In declaring my research agenda in Chapter One, I asked of the postage stamp: Is it a record of achievement, a target, or a didactic aspiration? It is all these things and more.

The scientists celebrated on postage stamps are largely local heroes, although there are a key group of universal heroes. Women are not well represented on postage stamps, which might reflect the fact that few women have historically been engaged in science.

However, with the growing sophistication of scientific images on stamps, the level of context has increased. Indeed, there are early examples where the achievement surpasses the need to recognise the scientist, because it was expected that people would know who was responsible. Examples of this are contained in the 1967 *Discovery and invention* set from Great Britain featuring radar (Watson-Watt), penicillin (Fleming), jet engines (Whittle) and television (Logie Baird) equipment. In the 1990s, when these achievements were revisited on stamps, the scientists' names are included in the text in addition to a more descriptive image. Such trends have been followed, reflecting how science communication practice has changed with time.

Coexisting with associating scientist and achievement, message context has become more comprehensive since the later 1990s. In one example, the wheel has turned full circle and some science issues have shown extremely detailed images of modern medical proceedures with a textual description of the name of the technique without mentioning the name of the inventor of the technique. A subsequent issue has used a similar image, and now names the leader of the team that has developed the proceedure. A balance is being achieved, and modern designs reflect the message that the postal administration is wanting to tell and the awareness of the designer of the science he is illustrating.

The postage stamp is an audience-based local medium containing a message made available through everyday use, which adds to the literacy of science communication. It is a means of disseminating information in an engaging manner, without any pressure, and may well promote public conversation. The move towards contextual images that seek to challenge or engage the public, the stamp as a lens, may well reflect the trend in science communication from a deficit model to models that seek to raise awareness or promote engagement with science.

Significance of the study

I have conducted a comparative study in time, (from the nineteenth century), but also in space, (across ten different countries), as was suggested for future research by Raento and Brunn (2005). From a science communication perspective, I have been able to show that there has been a greater emphasis on context on postage stamps from the mid-1990s. It is a date that makes sense, I believe, as it coincides with public take-up of the internet and an increase in content of the validity of the science message subjected to a more stringent scrutiny by the public. The millennium is another meaningful marker. Those countries that celebrated the event did so with different approaches to illustrate their messages, coinciding with the adoption of the public awareness communication model.

This first examination of the science messages on stamps has shown it to be a relevant addition to research initiatives in science communication, providing a platform for further study.

Developing the mirror or lens analysis on a stamp by stamp basis, although subjective and subject to understanding of the science perspective at the time of issue, has proved to be a strong indicator of how and why postal authorities have issued stamps. Has the image and context been developed as a lens to influence public behaviour, to get people thinking, or does it, as a mirror, reflect a reality with which the authority is comfortable? It is a technique that could apply to images across publishing.

Limitations of this study

It was practical to limit the number of countries to ten. Had I been able to extend the number I would have examined the stamps of Israel, as a politically-created country with a religious background, and India, whose scientific achievements are well recognised. I did also consider Japan and Brasil but, as discussed in the background to the study, preliminary work has been done on these countries, although not yet from a science perspective.

There is very limited research available on whether the message on the postage stamp has been read and caused any change in behaviour or understanding as a result of exposure to the message. Philip Parker of Royal Mail in London gave me copies of research Royal Mail had conducted, but these surveys, although of general interest, did not have a science focus.

Recommendations for further research

Any recommendations for further research will include the larger question of the public awareness of science and include the postage stamp as one of the miriad of sources from which information or data about science might be gleaned. The following questions arise directly from this study:

1. Are stamps contributing to a better understanding of the scientific problems that confront humanity today and of the measures to understand them?
2. Are the postal authorities creating celebrities through the scientists they feature on postage stamps, and to what effect?
3. Some countries issue stamps for purely commercial gain, the numbers of which are far in excess of any possible fiscal use. Some issues are not available in the home country's post offices. An analysis of the science stamps issued by these postal administrations, countries on the edge, as it were, could provide a quite different set of heroes.

Science on stamps

Although less standard mail is being sent, the issue of postage stamps and the opportunities to share messages with the public is expanding. The postage stamp or its equivalent will be raising the issues of science for a long time.

The underlying principles of science communication are reflected on postage stamps. The previous mode — a one-way informative educating image describing science — has shifted to a public awareness model. From 2000, stamp messages have moved away from being from information giving and towards provided context, so that data are illustrated without any obvious editorial bias.

The theory, practice and research in science communications is richer because of the existence of the postage stamp.

Bibliography

Altman, D. (1991) *Paper Ambassadors: The politics of stamps*. North Ryde: Angus & Robertson.

Australia Post (2011) 'Annual Report 2010–11'. Available at: http://auspost.com.au/media/documents/2010-11-integrated-annual-report-web.pdf.

Australian Dictionary of Biography (2012a) 'Dame Jean Macnamara'. Available at: http://adb.anu.edu.au/biography/macnamara-dame-annie-jean-7427.

Australian Dictionary of Biography (2012b) 'Howard Florey'. Available at: http://adb.anu.edu.au/biography/florey-howard-walter-10206.

Barr, T. (1993) *Reflections of Reality: The media in Australia*. Adelaide: Rigby.

Baudrillard, J. (1988) *Jean Baudrillard: Selected writings*, edited by M. Poster. Stanford: Stanford University Press.

Berglez, P. (2012) 'From Risk to Threat: Social representations of climate change in the media and among citizens'. Available at: http://www.aka.fi/Tiedostot/Tiedostot/ILMASTO/FICCA-työpajan_esitys%20Berglez.pdf.

Bowler, P. J. (2009) *Science for All*. London: University of Chicago Press.

Brooks, J. (1976) *Telephone: The first hundred years*. New York: Harper Collins.

Brown, S. E. (2008) *The New Zealand Collection 2008*. Wanganui: New Zealand Post Limited.

Bruce, R. V. (1990) *Alexander Graham Bell and the Conquest of Solitude*. New York: Cornell University Press.

Brunn, S. D. (2000) 'Stamp Iconography: Celebrating the independence of new European and Central States', *Geojournal*, Volume 52, Number 4, pp. 315–323.

Brunn, S. D. (2011) 'Stamps as Messengers of Political Transition', *The Geographical Review*, Volume 101, Number 1, pp. 19–36.

Burke, K. (2009) *The Stamp of Australia: The story of our mail from Second Fleet to twenty-first century*. Crows Nest: Allen & Unwin.

Catania, B. (1994) *Antonio Meucci: L'inventore e il suo tempo*. Rome: Seat.

Catania, B. (2002) 'The U.S. Government Versus Alexander Graham Bell: An important acknowledgement of Antonio Meucci', *Bulletin of Science, Technology and Society*, Volume 22, Number 6, pp. 426–442.

Cathcart, M. (2009) *The Water Dreamers*. Melbourne: Text Publishing.

Challoner, J. (2009) *1001 Inventions That Changed The World*. Hauppauge: Barron's Educational Series.

Chapman, K. (1994) 'Wedgewood Benn "PMG Extraordinary"', *British Philatelic Bulletin,* Volume 32, pp. 104–108.

chemistryexplained.com (2012) 'Justus von Liebig'. Available at: http://www.chemistryexplained.com/Kr-Ma/Liebig-Justus-von.html.

Child J. (2005) 'The Politics and Semiotics of the Smallest Icons of Popular Culture: Latin American stamps', *Latin American Research Review*, Volume 40, Number 1, pp. 108–137. Available at: https://muse.jhu.edu/login?auth=0&type=summary&url=/journals/latin_american_research_review/v040/40.1child.html.

Child, J. (2008) *Miniature Messages: The semiotics and politics of Latin American postage stamps*. Durham: Duke University Press.

Chubb, I. (2011) 'Does Australia Care About Science?', The Conversation. Available at: http://theconversation.com/does-australia-care-about-science-4011.

Collins, H. M. and T. Pinch (1998) *The Golem: What you should know about science*. Cambridge: Cambridge University Press.

Darby, A. (2010) 'China Flags in Antarctic Intent', *Sydney Morning Herald*. Available at: http://www.smh.com.au/opinion/politics/china-flags-its-antarctic-intent-20100111-m287.html#ixzz21JE1mHz0.

Davies, J. (1999) *Royal Mail Millennium Stamps 1999*, Volume 16. London: Royal Mail.

Davies, J. (2000) *Royal Mail Millennium Stamps 2000,* Volume 17. London: Royal Mail.

Davies, J. (2010) *The Big Picture, Royal Mail Special Stamps 2010*. London: Royal Mail.

Davies, N. (2005) *God's Playground: A history of Poland*, 2 Volumes. Oxford: Oxford University Press.

Deering, J. (2011) *The UK Scene,* Volume 48, Number 2.

Dennison, B. (2010) 'History of Science Communication', paper presented at Latornell Conference, Ontario, Canada. Available at: http://ian.umces.edu/blog/2010/12/26/bill-dennison-speech-to-latornell-conference-ontario.

Department of Industry (2010) *Inspiring Australia: A national strategy for engagement with the sciences.* Available at: www.innovation.gov.au/InspiringAustralia.

De Young, G. (1986) 'Postage Stamps and the Popular Iconography of Science', *Journal of American Culture,* Volume 9, Number 3, pp. 1–14.

Di Somma, M. (1999) *New Zealand Leading the Way: Creativity and versatility.* Auckland: New Zealand Post.

Di Somma, M. (2000) *The Millennium Album.* Auckland: New Zealand Post.

Eckersley, R. (2001) 'Postmodern Science: The decline or liberation of science', in S. M. Stocklmayer, M. M. Gore, C. R. Bryant (eds), *Science Communication in Theory and Practice.* Dordrecht: Klewer Academic Publishers.

Ekker, C. (1969) 'Stamps as Unique Primary Research Materials', *Topical Time* Volume 20, Number 5, pp. 40–41.

eduplace.com (2012) 'Vitus Bering 1681–1741'. Available at: http://www.eduplace.com/kids/socsci/ca/books/bkd/biographies/bk_template.jsp?name=beringv&bk=bkd&authorname=beringv&state=ca.

Encyclopaedia Britannica (2012). 'pitchblende'. Available at: http://www.britannica.com/EBchecked/topic/462007/pitchblende.

Encyclopedia Britannica (2013a) 'al-Burini'. Available at: http://www.britannica.com/EBchecked/topic/66790/al-Biruni.

Encyclopedia Britannica (2013b) 'Avicenna'. Available at: http://www.britannica.com/EBchecked/topic/45755/Avicenna.

Ericksen *et al.* (2008) *Collection of Australian Stamps '08.* Melbourne: Australian Postal Corporation.

Ericksen *et al.* (2009) *Collection of Australian Stamps '09* Melbourne: Australian Postal Corporation.

European Society for the History of Science (2011) Visual representations in science. Available at: http://www.eshs.org/content/779.

Fahour, A. (2012) 'Message from the Managing Director', Australia Post Press Release.

Fiske, J. (1989) *Understanding Popular Culture.* London: Routledge Company.

Flude, A. (2001) 'Exploration and Settlement'. Available at: http://homepages.ihug.co.nz/~tonyf/explore/explore.html.

foxnews.com (2008) 'Stamp Honouring Biochemist Bears Error'. Available at: http://www.foxnews.com/wires/2008Jan15/0,4670,StampChemical,00.html.

Frewer D. (2002) 'Japanese Postage Stamps as Social Agents: Some anthropolical perspectives', *Japan Forum*, Volume 14, Number 1, pp. 1–19. DOI: 10.1080/09555800120109005.

Friedel, R. D., P. Israel and B. S. Finn (1986) *Edison's Electric Light: Biography of an invention*. New Brunswick: Rutgers University Press.

Furukawa, A. (1994) *Medical History Through Postage Stamps*. St. Louis: Ishiyaku EuroAmerica.

Garmonsway, G. N. (ed.) (1965) *The Penguin English Dictionary*. Harmondsworth: Penguin.

Gatrell, P. (1994) *Government, Industry and Rearmament in Russia: 1900–1914*. Cambridge: Cambridge University Press.

Gentleman, D. (1972) *Design in Miniature*. London: Watson-Guptill.

Gregory, J. and S. Miller. (1998) *Science in Public: Communication, culture and credibility*. New York: Plenum Press.

grist.org (2005) 'French Constitution Gets a Dash of Green'. Available at: http://grist.org/article/case-france/.

Hamilton-Bowen, R. (2009) *Hibernian Catalogue of the Postage Stamps of Ireland*. Rodgau: Roy Hamilton-Bowen, BPP.

Harding, S. (ed.) (2011) *The Postcolonial Science and Technology Studies Reader*. London: Duke University Press.

Haskins, E. V. (2003) '"Put your stamp on history": The USPS commemorative program *Celebrate the Century* and postmodern collective memory', *Quarterly Journal of Speech*, Volume 89, Number 1, pp. 1–18.

Horne, D. (1986) *The Public Culture*. London: Pluto.

Hughes, T. P. (1977) 'Edison's Method', in W. B. Pickett (ed.), *Technology at the Turning Point*. San Franscisco: San Francisco Press.

Hymans, J. (2004) 'The Changing Colour of Money: European currency iconography and collective identity', *European Journal of International Relations*, Volume 10, Number 1, pp. 5–32.

Jeffries, H. (2011) 'Living figures to be shown on US stamps', *Gibbons Stamp Monthly*, Volume 42, Number 6.

Jennings, P. (2012) 'The People's Stamps: First past the post!' *Gibbons Stamp Monthly*, Volume 43, Number 5.

Jones, R. A. (2001) 'Heroes of the Nation?: The celebration of scientists on the postage stamps of Great Britain, France and West Germany', *Journal of Contemporary History*, Volume 36, Number 3, pp. 403–422.

Jones R. A. (2004) 'Science in National Cultures: The message of postage stamps', *The Public Understanding of Science*, Volume 13, Number 1, pp. 75–81.

Jowett G. S. and V. O'Donnell (2012) *Propaganda and Persuasion*, fifth edition. London: Sage.

Kang, D. and A. Segal (2006) 'The Siren Song of Technonationalism', *The Far Eastern Economic Review*, Volume 169, Number 2, pp. 5–11.

Kemp, M. (2006) 'Stamping his Authority', *Nature*, Volume 439, Number 7075, p. 396.

Kennedy, M. (2007) *Royal Mail Special Stamps 2007*. London: Royal Mail.

Kevane, M. (2006) 'Official Representations of the Nation: Comparing the postage stamps of Sudan and Burkina Faso'. Available at: http://ssrn.com/abstract=1115505.

Kress, G. R. and T. Van Leeuwen (2006) *Reading Images: The grammar of visual design,* second edition. New York: Routledge.

Lehmkuhl, M., C. Karamanidou, B. Trench, Tuomo Mörä and K. Petkova (2011) 'Science in Audiovisual Media: Production and perception in Europe'. Available at http://www.polsoz.fu-berlin.de/en/kommwiss/v/avsa/Downloads/finalreport_avsa_2010.pdf.

Lehmkuhl, M., C. Karamanidou, T. Mörä, K. Petkova and B. Trench (2012) 'Scheduling Science on Television: A Comparative analysis of the representations of science in 11 European countries', *Public Understanding of Science*, Volume 21, pp. 1002–1018.

Leon, B. (2004) 'Science Popularisation through Television Documentaries: A study of the work of British wildlife filmmaker David Attenborough'. Paper presented at the 5th International Conference of Science and Technology. Available at: http://www.pantaneto.co.uk/issue15/leon.htm.

Library and Archives Canada (2012) Dr. Norman Bethune. Available at: http://www.collectionscanada.gc.ca/physicians/030002-2100-e.html.

Liew J. (2009) 'Stamp Collecting Enjoys a Surge in Popularity', *The Telegraph*. Available at: http://www.telegraph.co.uk/news/newstopics/howaboutthat/6753169/Stamp-collecting-enjoys-surge-in-popularity.html.

Linn's Stamp News (2012) Available at http://www.linns.com/.

Lorimer, R. and P. Scannell. (1994) *Mass Communications: A comparative introduction*. Manchester: Manchester University Press.

Mackay, J. (2011) *The Complete Guide to Stamps and Stamp Collecting*. Wigston: Anness.

Mackay, J. A. (1976) *Encyclopaedia of World Stamps, 1945–1975*. London: Lionel Leventhal.

Marshall, P. D. (1997) *Celebrity and Power: Fame in contemporary culture*. Minnesota: University of Minnesota Press.

marxists.org (2012) 'Selected Works of Mao Zedung: In memory of Norman Bethune'. Available at: http://www.marxists.org/reference/archive/mao/selected-works/volume-2/mswv2_25.htm.

Masters, I. (2012) Head of Stamps, NZ Post, personal communication during meeting on March 16, Wellington, New Zealand.

McAllister, B. (2012) 'Governors Show No Interest in Supporting Living Person Stamp', *Linn's Stamp News*, Volume 85, Number 6.

McCalman, I. (2009) *Darwin's Armada: How four voyagers to the Australasia won the battle for evolution and changed the world*. Camberwell: Penguin.

McQuail, D. (1994) *Mass Communication Theory: An introduction*, third edition. Thousand Oaks: Sage Publications.

Merton, R. K. (1957) 'Priorities in Scientific Discovery: A chapter in the sociology of science', *American Sociological Review*, Volume 22, Number 6, pp. 635–659.

Michaud, J. (2010) 'The Apparatus of Democracy', *The New Yorker*. Available at: http://www.newyorker.com/online/blogs/backissues/2010/09/the-apparatus-of-democracy.html.

Miller, J. (2011) 'Images of Mathematicians on Postage Stamps'. Available at: http://jeff560.tripod.com/stamps.html.

Moore, A. L. (2003) *Postal Propaganda of the Third Reich*. Atglen: Schiffer Publishing.

Moyal, A. (2012) 'Owen Stanley and the Rattlesnake', *The National Library Magazine*, Volume 4, Number 2, pp. 8-11.

National Academy of Sciences (2005) 'International Geophysical Year'. Available at: www.nas.edu/history/igy.

National Center for Biotechnology Information (2013) 'Rudolf Stefan Weigl – Scientist and Human Being'. Available at: www.ncbi.nlm.nih.gov/pubmed/12926332.

New York Times (1882) 'A New Incandescent Light: A German electrician's invention', *New York Times*, 30 April 1882. Available at: http://query.nytimes.com/mem/archive-free/pdf?res=FB0B17F83A5A11738DDDA90B94DC405B8284F0D3.

Nielson, M. (2012) *Reinventing Discovery*. Woodstock: Priceton University Press.

Nisbet, M. C. and D. A. Scheufele (2009) 'What's Next for Science Communication?: Promising directions and lingering distractions'. *American Journal of Botany*, Volume 96, Number 10, pp. 1767–1778.

Nobel Foundation (1967) *Nobel Lectures, Physiology or Medicine 1901–1921*. New York: Elsevier.

Nobel Foundation (1967a) *Nobel Lectures, Physics 1901–1921*. New York: Elsevier.

Nora P. (1984) *Les Lieux de Mémoire*. Paris: Gallimard.

Novikov, S. (2012) 'Mstislav Vsevolodovich Keldysh', *MacTutor History of Mathematics*. Available at: http://www-history.mcs.st-andrews.ac.uk/Biographies/Keldysh_Mstislav.html.

Office of Science and Technology and the Wellcome Trust (2000) 'Science and the Public: A review of science communication and public attitudes to science in Britain'. Available at: http://www.wellcome.ac.uk/stellent/groups/corporatesite/@msh_peda/documents/web_document/wtd003419.pdf.

O'Sullivan C. J. (1988) 'Impressions of Irish and South African National Identity on Government Issued Postage Stamps', *Eire-Ireland*, Volume 23, pp. 104–115.

Palmer, S. (2012) 'Hegel's Owl'. Paper presented at the U3A ACT, Historians Talking About History Conference.

Parker, P. (2004) 'Stamps and philately', Royal Mail Conference Strategic Policy Document. London: Royal Mail.

Paterson, W. (2009) Postal Services: Global meltdown?, *Campbell Paterson Newsletter*, Volume 61, Number 3, pp. 2–5.

Paterson, W. (2012) *Campbell Paterson Catalogue of New Zealand Stamps*. Auckland: Campbell Paterson Limited.

Peirce, C. S. (1867) *On a New List of Categories*. Cambridge: American Academy of Arts and Sciences.

Pellechia, M. (1997) 'Trends in Science Coverage: A content analysis of three US newspapers', *Public Understanding of Science*, Volume 6, Number 1, pp. 49–68.

Petress, K.C. (1991) 'Postage Stamps as Rhetorical National Images', paper presented to the Southern States Communication Association Tampa, Florida. Available at: http://www.coursehero.com/file/1332874/ArticleH04/.

Philatelic Bulletin (1999) '1999 Favourite Stamps Poll', *British Philatelic Bulletin*, Volume 37, pp. 116–117.

Philatelic Bulletin (2000) 'Results of stamp poll 1999', *British Philatelic Bulletin*, Volume 38, pp. 116–117.

Philatelic Bulletin (2001) 'Results of stamp poll 2000', *British Philatelic Bulletin*, Volume 37, pp. 238–239.

Philatelic Bulletin (2010) 'Life-changing science', *Philatelic Bulletin*, Volume 47, Number 11, pp. 348–349.

Philsci, Philosophy of Science Association. 20th Biennial Meeting (Vancouver), 2006 Visual representations in science (abstract), contributed by W. M. Goodwin. Available at philsci-archive.pitt.edu.

Priest, S. H. (2009) 'Reinterpreting the Audiences for Media Messages about Science', in R. Holliman, E. Whitelegg, E. Scanlon, S. Smidt, and J. Thomas (eds), *Investigating Science Communication in the Information Age*. Oxford: Oxford University Press.

Raento, P. (2006) Communicating Geopolitics Through Postage Stamps: The case of Finland'. *Geopolitics*, Volume 11, Number 3, pp. 601–629.

Raento, P. and S. D. Brunn (2005) 'Visualising Finland: Postage stamps as political messengers', *Geografiska Annaler, Series B: Human Geography*, Volume 87, Number 2, pp. 145–163.

Rand-McNally (1977) *Encyclopedia of Transportation*. Chicago: Rand McNally.

Rees, M. (2010) 'The Science Citizen', Reith Lecture 2010. Available at: http://www.bbc.co.uk/programmes/b00sj9lh.

Reid, D. (1984) 'The Symbolism of Postage Stamps: A source for the historian', *Journal of Contemporary History*, Volume 19, Number 2, pp. 223–249.

Reid, D. (1993) 'The Postage Stamp: A window on Saddam Hussein's Iraq', *Middle East Journal*, Volume 47, Number 2, pp. 77–89.

Repcheck, J. (2007) *Copernicus' Secret*. New York: Simon & Schuster.

Riesch, H. (2011) 'Changing News: Re-adjusting science studies to on-line newspapers', *Public Understanding of Science*, Volume 20, Number 6, pp. 771–777.

Rodliffe, G. (2008) *Richard Pearce: Pioneer Aviator*. Auckland: Museum of Transport and Technology.

Rose, G. (2012) *Visual Methodologies*, third edition. London: Sage.

Rose S. (1980) *Royal Mail Stamps: A survey of British stamp design*. Oxford: Phaidon.

Rosenberg, M. (2012) 'Captain James Cook: The Geographic Adventures of Captain Cook — 1728–1779', *about.com*. Available at: http://geography.about.com/cs/captaincook/a/jamescook.htm.

Royal Mail editorial (2012) 'Royal Mail: The development of a stamp issue'. Available at: www.royalmail.co.uk.

Royal Society ad hoc Group (1985) 'The Public Understanding of Science'. Available at: https://royalsociety.org/~/media/Royal_Society_Content/policy/publications/1985/10700.pdf.

Schroeder, H., M. T. Boykoff and L. Spiers (2012) 'Equity and State Representations in Climate Negotiations', *Nature Climate Change*, Volume 2, Number 12, pp. 834–836.

Scott D. (1991) 'Posting Messages', *GPA Irish Arts Review Yearbook, 1990–1991*, pp.188-196.

Scott D. (1995) *European Stamp Design: A Semiotic Approach to Designing Messages*. London: Academy Editions.

Scott D. (2002) 'The Semiotics of the *Lieu de Mémoire*: The postage stamp as a site of cultural memory', *Semiotica*, Volume 2002, Number 142, pp. 107–124.

Scott Carlton, R. (ed.) (1997) *International Encyclopaedic Dictionary of Philately*. Iola: Krasue Publications.

Scott Publishing Company (2009) 'Scott Standard Postage Stamp Catalogue', Sidney, Ohio: Amos Hobby Publishing.

Searle, S. D. (2011) 'Scientists' Communication with the General Public: An Australian survey', unpublished PhD thesis, The Australian National University, Canberra. Available at https://digitalcollections.anu.edu.au/handle/1885/8973.

Sergey, M. and D. Kazbek (2013) 'Mstislav Keldysh', *Philamath*, Volume 34, Number 3, pp. 3–5.

Shackleton, T. (1994) *Royal Mail Special Stamps 1994*. London: Royal Mail.

Stanley Gibbons Publications (2002) *Stanley Gibbons Stamp Catalogue Part 5: Czechoslovakia and Poland*. Ringwood: Stanley Gibbons Publications Ltd.

Stanley Gibbons Publications (2006) *Stanley Gibbons Stamp Catalogue Part 6: France*. Ringwood: Stanley Gibbons Publications Ltd.

Stanley Gibbons Publications (2007) *Stanley Gibbons Stamp Catalogue Part 7: Germany*. Ringwood: Stanley Gibbons Publications Ltd.

Stanley Gibbons Publications (2008) *Stanley Gibbons Stamp Catalogue Part 10: Russia*. Ringwood: Stanley Gibbons Publications Ltd.

Stanley Gibbons Publications (2010) *Stanley Gibbons Stamp Catalogue Part 10: Russia*. Ringwood: Stanley Gibbons Publications Ltd.

Stocklmayer, S. M. (2013) *Communications and Engagement with Science and Technology*. London and New York: Routledge.

Stoetzer, C. (1953) *Postage Stamps as Propaganda*. Washington: Public Affairs Press.

Sweet, F. (2003) *Royal Mail Special Stamps 2003*, Volume 20. London: Royal Mail.

Tanner, A. M. (1894) 'The Goebel-Munchhausen Lamp Story', *The Electrical Review*, Volume 34, Number 845, pp. 113.

Taylor, M. (2012) 'Framing in the Context of Science Communication', Unpublished PhD Thesis, The Australian National University.

techdirections.com (2011) 'Heroes of Science and Technology Posters'. Available at: http://www.techdirections.com/postersscience.html.

Terzian and Grunzke (2007) 'Scrambled Eggheads', *Public Understanding of Science*, Volume 16, Number 8, pp. 407–419.

Thomas, M. (2012) 'A Land of Cold Truths', *The Canberra Times*, 1 December.

Tink, A. (2009) *William Charles Wentworth: Australia's greatest native son*. Crows Nest: Allen & Unwin.

Trammell J. (2011) 'Stamps as Cultural Artifacts', *Stanley Gibbons Monthly*, Volume 42, Number 1, pp. 98–99.

Tsou, T. (1968) *The Cultural Revolution and Post-Mao Reforms: A historical perspective*. Chicago: University of Chicago Press.

United Kingdom Foreign and Commonwealth Office (2012) 'Country Profile: Russia'. Available at: https://www.gov.uk/foreign-travel-advice/russia.

University of Frankfurt (2011) 'Physics-Related Stamps'. Available at: http://th.physik.uni-frankfurt.de/~jr/physstamps.html.

Vickery, B. C. (2000) *Scientific Communication in History*. Maryland: Scarecrow Press.

Vyvyan, R. N. (1974) *Marconi and Wireless*. Leeds: E P Publishing.

Waldmann, W. (ed.) (2002) *Encyclopedia of World Explorers*. Connecticut: Easton Press.

Weber R. L. (1980) *Physics on Stamps*. London: The Tantivy Press.

Wilford, J. N. (2012) '50 Years Later, Celebrating John Glenn's Feat', *The New York Times*. Available at: http://www.nytimes.com/2012/02/14/science/space/50-years-later-celebrating-john-glenns-great-feat.html?pagewanted=all&_r=0.

Wilson R. J. (2001) *Stamping Through Mathematics*. New York: Springer-Verlag.

Wood, M., B. Cole and A. Gealt (1989) *Art of the Western World: From Ancient Greece to Post-Modernism*. New York: Simon & Schuster.

Yvert et Tellier. (2010) *Le Petit Yvert: Catalogues des Timbres-Poste De France*. Paris: Yvert et Tellier.

Zsolt, M. (2012) Manager, Philatelic Services, Australia Post. Personal communication.

www.ingramcontent.com/pod-product-compliance
Lightning Source LLC
Chambersburg PA
CBHW061130010526

44116CB00025B/2992